Illness makes a powerful argument for exploring the experience of illness and the associated philosophical questions. Carel's inclusion of herself in the book is often moving and shows well the power of bringing philosophy and personal life together."
Philosophy in Review

"This book offers an important contribution to the ongoing project of the phenomenology of illness, and offers a powerful argument for the inclusion of applied phenomenology in medical and healthcare training. One of the main strengths of this book is that it forces you to think, and to think philosophically. Carel neatly lifts philosophy off the page, and places it out there like a talisman in our everyday life. The book deserves to be read widely by the public, and I would suggest needs to be read widely by clinical practitioners as a point of reference for their own practice."

Metapsychology

"*Illness* offers us something that we all need to read and think about ... If I were to write a book about illness, I would want it to be just like this one."

Arena

"a marvelous book ... a very clear and detailed account of the phenomenology of illness and the contribution it could make to medical practice and research."

Homeopathy

"A masterpiece. Moving seamlessly between an unsparingly honest personal narrative and philosophical reflections on our condition as embodied subjects, Havi Carel has fashioned a uniquely authentic account of the lived experience of illness. It should be read – and re-read – by everyone who is professionally involved with illness, who is ill, or is likely to become ill; which is to say, by all of us."

RAYMOND TALLIS, *Fellow of the Academy of Medical Sciences, and formerly Professor of Geriatric Medicine, University of Manchester*

"A genuinely important philosophical work. Carel succeeds in offering a wide-ranging, original, wholly convincing and quite beautiful account of the phenomenology of illness. This is a remarkably insightful book about what it is to be human and how to live. Anybody who cares about who they are and how they live ought to read it."

MATTHEW RATCLIFFE, *Professor of Philosophy, University of Durham*

"A tremendous achievement, as well as being a very moving personal document."

CHRISTOPHER BERTRAM, *Professor of Social and Political Philosophy, University of Bristol*

Illness

The Cry of the Flesh

REVISED EDITION

Havi Carel

LONDON AND NEW YORK

First published in 2008 by Acumen
Revised edition published 2013

Published 2014 by Routledge
2 Park Square, Milton Park, Abingdon, Oxon OX14 4RN
711 Third Avenue, New York, NY 10017, USA

Routledge is an imprint of the Taylor & Francis Group, an informa business

ISBN 13: 978-1-84465-753-7 (pbk)

British Library Cataloguing-in-Publication Data
A catalogue record for this book is available from the British Library.

Typeset in Warnock Pro.
Printed and bound by CPI Group (UK) Ltd, Croydon, CR0 4YY

To Samir, for everything

Gracias a la vida que me ha dado tanto

Contents

Acknowledgements

This book was written during a period of research leave funded by the University of the West of England. It was reworked into a revised edition during a period of research leave awarded by my current employer, the University of Bristol, and funded by the British Academy. I am grateful for these periods of research leave.

Many people have read the book over the years in draft form and as a published text, and provided valuable feedback. I am not able to thank them all in person, but I have benefited greatly from exchanges with patients and their families, health professionals and fellow philosophers. In particular I would like to thank the following people. Eran Dorfman, an astute reader and supportive friend, pushed me to do (and write) as I say. Iain Grant provided valuable comments on an early draft, shared with me his insatiable appetite for philosophy. John Sellars made structural suggestions that greatly improved the book. Christopher Wakling read the manuscript and provided reassuring feedback. I am especially lucky to have had Samir Okasha scrutinize the text and provide the best possible feedback. I am greatly indebted to him for the time he invested, directly and indirectly, in this book.

A number of people provided comments on draft chapters: Michael Brady, Matthew Broome, Jordan Carel, Sari Carel and Sarah Dietz. Matthew Ratcliffe and Raymond Tallis reviewed the manuscript for Acumen. Thomas Baldwin suggested Buber's I–Thou relationship as a way of understanding the patient–physician interaction discussed in Chapter 2. I am grateful to Mark Vernon for commissioning the book and providing useful feedback along the way, and to Steven Gerrard of Acumen for supporting the project from the outset.

I am especially grateful to those entrusted with the unenviable job of looking after my health: the Respiratory Department at the Bristol Royal Infirmary, and especially my respiratory nurse, Ros Badman, and the other nurses who have supported me in all things big and small, over the years; my consultant, Dr Liz Gamble; the physiologists, who run my lung function tests. I'm very grateful to Fraser McKay of Air Liquide, for making deliveries through snow and ice and for taking patients seriously.

Thanks also to the Department of Cardiopulmonary Transplantation at the Freeman Hospital in Newcastle; and to Professor Simon Johnson, Director of the National Centre for LAM at Queen's Medical Centre in Nottingham, the best physician patients could ask for. His remarkable ability to understand illness without being ill himself has made all the difference to me. Thanks also to Jan Johnson, the vital core of LAM Action, who has been part of this organization since its inception and has given many newly diagnosed patients their first support. I also thank those on the LAM Action executive committee, especially Anne Tattersfield and Jane Tallents, for supporting LAM patients over the years. Finally, I thank my fellow LAM patients for many years of mutual support and sharing of advice and information.

I would like to thank Editions Rodopi for allowing me to use material from my monograph, *Life and Death in Freud and Heidegger* (2006) in Chapter 4; and Springer for allowing me to use material published in "Can I be Ill and Happy?", *Philosophia* 35(2) (2007), 95–110, in Chapter 3.

Finally, I feel very lucky to be able to thank my family. My mother and father, Cynthia and Rafael Carel, have been a pillar of strength and illumination to me; they have given me more than I could have hoped to ever receive. My beautiful sister, Sari, brought many magical things into existence, not least my niece, Petra. Sari has been my closest ally since we were terribly dressed teenagers in the 1980s. My brother, Jordan, has been everything a big brother should be, with his explosive intelligence and excellent pillow-fighting skills. My son, Sol, has brought elation and completeness into my life, just as I was giving up. And lastly, I thank my husband, Samir Okasha, for his kindness, tenacity and optimism. Together we succeeded in the difficult task of turning a disaster into an obstacle. This book is dedicated to him, with love and gratitude far greater than can be conveyed by black marks on a white page.

Havi Carel
Bristol

Preface to the revised edition

When I was diagnosed with a lung condition in 2006, my personal life and my life as a philosopher were split apart. The effect of the diagnosis on my personal life was enormous. I was told that my prognosis was poor (happily, that is no longer the case) and that I would have to use oxygen and possibly have a lung transplant. I was also told that I would be unable to have children. Everything I had until that point taken for granted was thrown into the air. I experienced extreme uncertainty and anxiety. I had to adjust to, and make sense of, a huge shift in my life plans and expectations and my day-to-day routine. "What does philosophy have to do with all this?", I asked. At the time, I felt alienated from philosophy and angry at its abstract nature. My concerns were highly personal and idiosyncratic, but I nonetheless felt a demand placed on me as a philosopher to find concrete ways in which philosophy could help me cope with my illness. So I decided to take philosophy to task and see how philosophy might be used in the context of illness.

When I started reading the literature in the philosophy of medicine – where I thought I would find detailed analyses of the experience of illness – I realized that there is a serious oversight within mainstream philosophy, a failure to address the thing that matters most to ill people: how they feel, what they experience, how illness changes their lives. With the exception of a few authors such as S. Kay Toombs, Matthew Ratcliffe and Fredrik Svenaeus, the phenomenology of illness was largely untreated by philosophers. I thought that since illness is a profound and universal experience shared by almost all humans, philosophy ought to study it. However, when I started looking at the literature, I discovered that although philosophers have written much about death, illness has not received the same systematic treatment. At first I wondered if there was anything of distinctly *philosophical* value in illness. I discovered there was. Illness changes not only the content of experience (by limiting it, or making it painful) but also its structure. The experience of space and time, for example, is modified in many somatic and mental disorders (as discussed by Ratcliffe and Thomas Fuchs). Illness also calls into question one's values and sense of fairness and disrupts the meaning-making we routinely engage in. I thought that given its dramatic impact on virtually all domains of life, illness would naturally lend itself to philosophical analysis. But within philosophy very little was written on illness from this perspective. To address this lacuna I took some ideas from phenomenology (mainly from Merleau-Ponty and Heidegger) and wrote a philosophical analysis of the experience of illness, which combined my experiences with philosophical reflection. This account was published as the first edition of this book, *Illness*, in 2008.

The book triggered responses from health professionals, patients and their families, policy-makers and medical educators. These responses, in turn, got me involved in interesting activities, both within academia and outside it. I took part in empirical studies of illness, advised professional bodies, represented patients in meetings, participated in a meta-ethnography of medical intervention and presented my work to health professionals and medical educators. I was delighted to be able to engage with these diverse groups but also worried. How can I translate arcane phenomenological terminology to non-philosophers? Would the ideas seem too abstract? How can I make philosophy useful in the most immediate and direct sense to these people, many of whom had no previous contact with philosophy? These were some of the challenges I faced in trying to import philosophy, and in particular phenomenology, from academic settings to – literally – bedside conversations.

I also faced the challenge of bridging different worlds and different realities: those of philosophers, medics, patients and family members, each of whom has their own view of illness. It is not enough, I quickly learned, to get people together in a room. How do I ensure that the practical usefulness of philosophy is conveyed to them and that they understand and engage with each other? Much of the initial work in bringing together philosophers, practitioners, service-users and members of the public, I can say in hindsight, is a work of translation and integration. The first step is to develop a common ground and a compassionate space of intellectual curiosity and openness between speakers coming from completely different backgrounds. I learned that there is an inevitable tension and perhaps even suspicion in some of these encounters and that space must be made for hesitant openings, jerky attempts at communication, and incomprehension, while this shared ground is slowly established.

Many things that I took for granted as a philosopher – a certain style of enquiry, attention to detail, an appetite for the abstract – were foreign to some of my interlocutors. For example, by teaching medical students I came to learn that these students often read philosophical texts with a healthy dose of frustration and incomprehension. "What does this article say?" and "How will this be useful to me in my future career as a doctor?" were common questions from students that called upon me better to understand their conception of learning. Taking them on a philosophical journey, when the vast majority of their study consists of learning, not questioning, facts, was often challenging. The idea of open-ended reflection, sustained debate and different points of view was foreign to many of these students and demanded a particular pedagogical effort on my part.

Despite these challenges, I managed to apply philosophy, and in particular phenomenology, in the healthcare arena. Here are a few examples of the work that followed on from *Illness*. I developed a workshop for health professionals called "What is Illness?". This two-hour workshop invites health professionals to engage with some philosophical views on health, illness and disease. The workshop first contrasts the naturalistic approach (which sees disease as a value-free concept, as biological dysfunction) with the "normativist" approach, which sees disease as a value-laden term. Participants are then invited to discuss the consequences of each approach to actual healthcare provision. The second part of the workshop presents the phenomenological approach to illness and asks participants to consider how a study of the experience of illness as lived by the ill person may inform their professional practice. What would change in how they administer care to patients? Would a hospital be differently designed? Would a ward round be conducted in a more participative manner? For many health professionals taking part in the workshops engaging with abstract questions was a new experience. Some of them, I think, were not convinced that this was time wisely spent. When I suggested the workshop to someone in charge of continuing professional development of nurses, she said "We have to do hand hygiene and practical stuff. Your work is just icing on a cake. We don't have time for cake." Through these workshops I learned that the abstract discussion was not enough in this context and that I needed to make a bigger effort to connect the theoretical ideas to health professionals' practical concerns and concrete practices.

I then began to share the ideas discussed in *Illness* with different audiences. I spoke to the British Pain Society about the experience of pain. I discussed somatic and mental illness with psychiatrists and respiratory physicians. I took part in "Medicine Unboxed" and "Medfest", and was the first patient to present a "Schwartz Round", in which health professionals heard my account of being given a serious diagnosis. A new area I am now working on (with Ian James Kidd) views illness as epistemic injustice, working out how testimonial injustice and, more seriously, hermeneutic injustice play a role in the interaction between health professionals and patients and how phenomenology may help to ameliorate these kinds of injustice suffered by ill people.

Thinking more about the application of phenomenology to illness also led me to develop some of the themes of *Illness* into a practical tool: a "patient toolkit" that draws on philosophical resources to support patients. This idea was developed in a paper that was published in the *Journal of Medicine and Philosophy* in 2012. The paper suggests that phenomenology could prove useful for patients, giving them tools to reflect on and expand their understanding of their illness. The idea is to offer philosophical patient support that does not take an already developed model of illness and present it to patients. Rather, its aim is to provide a flexible individual tool that patients can use to develop their understanding of their illness. I hoped that such a phenomenological resource could help patients to philosophically examine their illness, its meaning, and its impact on their lives. The toolkit includes three steps, aimed at supporting patients who wish to examine and develop their understanding of illness.

The first step invites patients to see illness as a form of phenomenological reduction. Phenomenology is committed to making explicit aspects of experience that are overlooked by other approaches and may be poorly understood. An adequate approach to the experience of illness requires what Husserl calls the phenomenological reduction: a suspension of a "natural attitude" of implicitly accepting the background sense of belonging to a world and various interpretive dogmas along with it. Bracketing the natural attitude is a withdrawal from the ordinarily implicit commitment to the reality of the world. This involves setting aside the host of assumptions we normally have about the reality of a disease entity in order to examine the illness experience without reducing it to a disease process. Bracketing the natural attitude towards illness suspends the belief in the reality of an objective disease entity. This suspension does not deny the objective reality of disease processes, but shifting the focus away from the disease entity and towards the experience of it can disclose new features of this experience. We usually take the disease entity for granted and posit it as the source of the illness experience. But in fact, for the ill person, the illness experience comes before the objective disease entity. Once the belief in the objective disease entity is bracketed and we are distanced from our usual way of experiencing, we can begin to explore how illness appears to the ill person, its structure and its essential features.

The second step of the toolkit invites patients to thematize illness. "Thematizing" refers to the act of attending to a phenomenon, which makes particular aspects of it explicit. A theme for a particular consciousness is that upon which it focuses its attention. But this does not simply denote the intentional object. It also takes into account the kind of attentional focus given to an entity. Thematizing may include attending to the cognitive, emotive, moral or aesthetic aspects of a phenomenon. A patient may thematize her illness as a central feature of her life, attending to her symptoms as pervasive, while the physician may thematize the illness as a "case of cancer", attending to symptoms as diagnostic clues (Toombs has also written on this issue). The understanding that illness is not an objective entity and the exercise of thematizing may help patients because it enables moving away from prescriptive pronouncements toward a descriptive mode. In this step patients explore how illness may appear to other patients, physicians, family members and so on, and how illness may be viewed in different ways: evaluatively, practically, emotionally, morally and even aesthetically. By holding illness constant but changing the ways in which it is perceived and the perspectives of those experiencing it, we can gain a more comprehensive understanding of illness. Thematizing creates a complex, shifting, view of illness as moving from foreground to background, as changing in meaning and as consisting of multiple perspectives.

The third step of the toolkit is to take the new understanding of illness (as a form of distancing that has been thematized) and examine how it changes one's being in the world. The term "being in the world" is used by Heidegger to denote the human being in the broadest sense. Being in the world includes the biological entity, the person, and her environment and meaningful connections. The third step calls on participants to examine the ways in which illness has altered their being in the world. By moving away from a narrow understanding of illness as a biological process, a fuller account of illness as a new way of being in the world can be developed by patients.

In 2012–13, with the support of a British Academy fellowship and with two wonderful collaborators – Catherine Lamont, an artist, and Dr Louise Younie, a GP and medical educator – I shared the toolkit with a patient group. In several sessions they discussed the steps of the toolkit and how useful such a toolkit might be. Some of the benefits they mentioned are "being able to speak your mind" and having an attitude other than self-pity towards their illness. They suggested that such a toolkit might help patients take charge of their treatment and health, and might help patients find words to express ideas and emotions that are difficult. They also suggested that creating such a group on a continuing basis may provide a safe space for grieving for lost health and freedom and for reinterpreting their illness as less negative. We next plan to have a toolkit session with a mixed group of patients and doctors, in which the two groups of participants might be able to learn more about one another's concerns, pressures and assumptions.

All of this work emerged from the first edition of *Illness*. The text was written as a private, therapeutic affair, and I never anticipated the positive reception it would get from fellow patients, health professionals and others. I was overwhelmed and delighted by the responses to it, which taught me that more work is needed to understand and communicate the experience of illness. This revised edition reflects the growing interest in the topic; the work that followed from the first edition shows how much of the experience of illness is still philosophically unexplored.

Medically, too, things have changed since 2006. A possible treatment for LAM, a drug called sirolimus, was identified and anecdotal evidence indicated that it may slow or stop disease progression in some LAM patients. I started taking the drug, initially "off-label" and now as a proven treatment for LAM (following the publication of the results of the MILES trial). I was lucky to be among the patients who responded to the drug, and was granted a lengthy (to this day and counting!) respite from continuous deterioration of my lung function. I once again had the stability and continuity that underpin any good life. I could rise above the quicksand of illness and breathe deeply: a normal life was possible once more. I also became a mother, against the odds, to a beautiful, spirited boy whose name, Sol, symbolizes the light and warmth he brought into our lives. Both medical stability and motherhood are to me nothing short of a miracle. But I am also deeply conscious of the precariousness of life and the extraordinary luck that brought about the two events. I remain acutely aware that it could have easily been otherwise.

The cry of the flesh: not to be hungry, not to be thirsty, not to be cold. For if someone has these things and is confident of having them in the future, he might contend even with Zeus for happiness.

(Epicurus, *Gnomologicum Vaticanum*, §33)

Introduction

The first time I realized I couldn't do something I felt surprise. It came as an insult, an affirmation of my limited existence. I had just moved back to the UK and gone to a circuit-training class. The class consisted of a number of exercises, including skipping, weights and push-ups. There were a handful of people in the class, including two slim girls. I was OK the first few rounds, but as we continued, I was less and less able to breathe. No, it wasn't my muscles or agility, or the weights; it was my lungs. I couldn't breathe. The instructor looked at me with pity. Unfit, he must have thought; lazy woman. I was slower and slower, did fewer repetitions in each round, and couldn't, for the life of me, skip. All the while the two slim girls were going strong, lifting, bending, jumping and skipping. I came out of that class feeling disappointed, beaten. Why could everyone else do this and I couldn't? Why was I so breathless?

That weekend I went to a spinning class. We rode on stationary bikes and added dumb-bell exercises to the cycling. The same inability struck me again. When we were sprinting I had to reduce the bike's resistance to zero. Even so I could barely breathe. Again, I left the gym feeling devastated. Why was I so unfit? By then a creeping suspicion entered my mind: something was wrong, very wrong. But my naive internet searches revealed nothing. I went to see my doctor, who ordered a breathing test and an X-ray. These began to reveal the true extent of my problem. My lung capacity was down to 47 per cent of the predicted average for my age and size. My lungs were grossly hyper-inflated, the X-ray showing a mysterious reticular structure. The radiologist recommended "further imaging". "What could it be?" I asked the doctor. "I don't know," she said. "I've never seen anything like this. You have the lungs of a sixty-year-old who's been smoking a pack a day since they were fifteen." "But I don't smoke," I said. "And I'm completely normal and I've always been so healthy. What could this be?" The doctor had no answers and referred me to a consultant at the local hospital. Two months later I saw her. She examined me thoroughly, taking the time to ask plenty of questions. At the end she gave her verdict: "unusual asthma". She prescribed some inhalers and asked me to plot my breathing on a chart to see if steroids and bronchodilators would help.

I felt relieved. Asthma: a common, treatable condition, I thought. The familiar sight of the blue inhaler reassured me. I had some doubts about my atypical asthma, since nothing seemed to trigger it and I never had any distinct attacks of narrowed airways. But I deeply wanted to believe this interpretation of my poor lung function. I also had my alpha 1-antitrypsin levels checked to rule out deficiency. When the result came back saying "normal" I cried with relief at the reception desk of the health centre. I used the inhalers and plotted my breathing as instructed. But the inhalers did not help. My peak-flow readings were slowly but steadily decreasing. It was winter. It was very cold. I found cycling to work harder and harder. And then I caught a cold. In the beginning it seemed like the usual runny-nose type of cold. Every day I would wake up thinking: today, today it must get better.

But each day it got a little worse, and two weeks into that cold I had to put away my bike and get the bus to work. Cycling was just too hard. I became dizzy going uphill and a new kind of breathlessness, severe, dizzying, nauseating breathlessness, more akin to suffocation than to panting, became a familiar feeling to me. I developed a cough, a chest infection, the first I had ever had. I coughed so much and lost so much weight that I began to worry. A grave fear shadowed my days. This couldn't be asthma. What sort of thing did I have?

My father, a medical doctor, was very concerned. He arranged for a chest CT scan for me at the medical centre he runs. I went there in the morning and was out in five minutes. I was on holiday and went to meet my good friend Eran in a cafe. We chatted and laughed, but when we walked together down the street I felt so breathless I had to ask him to slow down. It was all so odd. I met another friend in the afternoon. We went to buy sunglasses, then went to a cafe and talked about film. She was heavily pregnant and I hadn't seen her in a long time. We had plenty to talk about. My parents picked me up from the cafe and we drove back to the medical centre to collect the test results. My father told us to wait in the car; he would be back in five minutes.

When twenty minutes passed and he was not back a blinding panic overtook me. I knew that my deepest, unspoken fears were coming true, that a nightmare scenario was about to unfold. I knew it was happening. "I'm going in," I said to my mother. I left the car and rushed into the building. I walked back into the CT department I had left cheerfully in the morning. The receptionist tried to stop me. I pushed past her into the radiologist's office. My father was sitting at the radiologist's desk, his elbows on the table and his head resting on his hands. The lights were dimmed and the screen shone before me. My lungs were up there and the radiologist was pointing to them and talking to my father.

The radiologist turned to me, surprised and displeased to see me in his office, normally off-limits for patients. "So you are the patient," he said. "Do you know what's going on?" "No," I said, "what *is* going on?" "I'll let you read about it. Sit down." I stumbled into a chair and he handed me a diagnostic manual. The size of the book impressed me: big, authoritative, full of diseases. The book was turned to a specific page, the page of my disease. He pointed to a long word: lymphangioleiomyomatosis. What is this? I thought. What is happening to me? I skimmed the obtuse text: pneumathorax, chylous effusion, diffuse bullae ... The terms meant nothing to me. There was an eerie silence in the room. I had obviously interrupted something, intruded on their work. There were two medical note-takers sitting on either side of the radiologist's desk. They each had a desk-lamp illuminating their small territory: two small circles of light. His dark shadow loomed in the middle. Two young women flanked a senior radiologist. They both looked at me, intent, embarrassed.

I reached the bottom line. Prognosis: ten years from onset of symptoms. I sat there, the heavy book in my lap. Ten years, I thought. That would make me forty-five. Ten years. Pain and fear struck like a physical blow. It is difficult to describe the physicality of bad news. I remember looking at the room and feeling confused: it looked the same, while my life had been turned upside down. Make it stop, I thought. This is the wrong story. Someone come and fix it. Someone do something. The realization that everything was about to change, that a new era was about to begin, seared like burning oil on skin. It crushed me with invisible force. It is difficult to describe the pain and fear that descended on me at that moment. Now I cannot imagine my life without this pain and fear.

The radiologist was now talking to my father. The term "DD" (differential diagnosis) was repeated. He was not talking to me. Suddenly he turned to me. "Do you have light brown marks on your back?" he asked. "What?" I asked, confused. "What kind of marks? Why?" He did no reply. "Turn around," he said. With no warning he pulled up my jumper, a lovely, lavender jumper my sister had given me the previous day. He tugged at it impatiently and I could hear the seams creak and then give in to his violent pull. He looked at my back and muttered, "Nope, nothing there". I later found out that he was trying to establish whether I had sporadic LAM (which mainly affects the lungs) or the hereditary disease tuberous sclerosis complex, which may also affect other organs, including the skin. But at the time he did not explain that to me. I later repaired the diagnosis jumper, as I now call it, but I have never worn it since.

I remember feeling vindicated. Yes, I was sick. It was something serious and it had a name. It even appeared in the diagnostic manual that was sitting on my lap. I also felt smugly pragmatic when I asked the radiologist, "So what do I do?" I was thinking about a course of treatment, surgery, medication. I was thinking how there would be a rough patch but then I would get better. "I don't know," he laughed uncomfortably. "I only diagnose. I don't treat." Silence again. One of the note-takers turned to me and said, "It came as a shock to you, eh?" Her discomfort and his artificial jolliness made me realize things were much, much worse than I thought. He told his assistants to look on the internet for a treatment. My heart sank as I realized that he knew almost nothing about the disease. In his forty-year career he had seen only three cases of LAM. And now he was Googling it. The internet did not yield much new information. A novel, cruel thought dawned on me: my disease had no treatment. My mind went blank; it contained no thoughts and only one emotion – fear. I said the only thing I could think of: "Can someone get my mother".

That night my family closed ranks. We gathered in our family home, the house I grew up in. No one knew what to say or do. I couldn't eat, or sleep, or talk. My brother went to the car and got a DVD: *March of the Penguins*, of all films. We sat in front of the television, the penguins' stoic determination mocking our helplessness: what *are* we supposed to do now? We had no script. The hoarse cries of the Emperor penguins bellowed in my ears and still haunts me if I happen to hear it today. To me it was the sound of lament and devastation.

In the months that followed I went through a bewildering array of emotions. At first I was shocked, then depressed, sometimes relieved that I still had decent lung function, then disillusioned as my condition deteriorated. First came the inhalers: three different ones. Then came the oxygen: first portable cylinders, then the night-time supply. Then came a difficult period, winter again, during which I lost more and more abilities. Many things became impossible for me: walking uphill, walking and talking at the same time, doing *anything* and talking at the same time, running for the phone, going upstairs without stopping, carrying anything heavy. Any of these caused me severe breathlessness.

During that time, it seemed that every week my world was shrinking more and more. Every week I discovered, in a grotesque reversal of childhood development, yet another thing I could no longer do. I cancelled my gym subscription. I took the bus. I no longer tried to scale a hill. During a trip to Scotland my husband and others went up Cairn Gorm mountain. They got up early, packed their supplies and strapped on their hiking boots. I was left behind, to go up the funicular railway with the young children and the mums. How I wanted to be hiking that day. How I wanted to be able to spontaneously burst into a run, rough and tumble with my nephews, scale a hill without pausing for breath every ten metres.

But I was locked in my body, trapped by the feeble lungs, the impaired gas exchange, the pain in my chest, the fear of suffering a collapsed lung. I had to learn to smile and say "Why don't you go on ahead?" I had to learn to stop trying to keep up. I had to learn to ask for help from friends and sometimes from strangers. I learned the hard way. One day I went to work in a taxi and was planning to come back in one, because I had to work late. I left the oxygen behind thinking I could manage without it if I didn't need to walk very far. I was walking to the pub, perhaps 200 metres, with a colleague. I had to stop and rest; I had to ask him to carry my bag. I remember his astonishment, his awkwardness, as my disability revealed itself. He tried to be nice. We tried to joke and chat. I couldn't breathe.

A new life descended on me and I gradually acclimatized to it. The gradient of any walk became the most important factor and an electric bike replaced my beloved yellow racer. I learned to walk and talk more slowly, not to chew gum while walking and to be extra careful not to inhale anything, from dust to bits of food. I slowed down. I imagine that this is what it must be like to grow old: to gradually realize that as your body loses capacities your world shrinks too. Except that old people have decades to prepare for this. I was thirty-five at the time of my diagnosis.

From that day – 10 April 2006 – my life changed beyond recognition and yet remained the same. I learned more about my embodied existence, about people's attitudes towards illness and disability, about the inability to speak of important things, than I had in the preceding decades of my life. I found myself in an odd position: young, but in some ways old; healthy-looking on the outside but gravely ill inside. I have a future but it is darkened by the threat of death. I am now familiar with the twilight world of lung transplantation. It is not the average world of a woman in her mid-thirties.

I found that I had to reinvent my life. I had to give up some friendships. I had to learn to be tough on myself and sometimes rude to others. When strangers approached me, I learned to fend them off with a grumpy mutter. When people were about to embark on a series of unwanted questions, I learned to defend myself with a stern look and an impatient tone. I learned to rethink my aspirations and plans. I relinquished the sense of control over my life that I previously had. And more than anything, I learned to love what I still had. So much so, that many people who meet me do not believe I am ill at all: that is, until they watch me walking to the corner shop. I learned to graciously accept compliments about how healthy I appear; I learned to ask for help when none was offered. I learned that people will not know anything about the world of illness unless I tell them. I learned to cope, to surrender vanities. I adjusted.

I learned to live a Janus-headed life: young but old, healthy-looking but ill, happy but also incredibly sad.

Those were the experiences that led me to write this book. The experience of illness and its sweeping effect on every aspect of life shocked me into thinking about these issues. This is a book founded on my experience of living with a degenerative and potentially fatal illness. Because of my training as a philosopher, my experiences pushed me to reflect abstractly on health and illness: what these concepts mean and how best to understand them. But when I started my research I found that the language and concepts routinely used to describe illness are inappropriate, incomplete and often misleading. I became increasingly aware of the impoverished language used in the medical world I encountered, which, in turn, led me to suspect that an impoverished concept of illness lay in the background.

And so I began to think about the concepts of illness and health particularly as they are used within the medical world. A new research project emerged, which aimed to make philosophical sense of the concept of illness. This book is the outcome of this project.

WHY PHENOMENOLOGY?

My main discomfort with the orthodox concept of illness is that it originates in a *naturalistic* approach. Naturalism is a label for a broad spectrum of views saying, roughly, that natural or physical facts are sufficient to explain the human world. On a naturalistic view, illness can be exhaustively accounted for by physical facts alone. This description is objective (and objectifying), neutral and third-personal. Naturalistic descriptions of illness exclude the first person experience and the changes to a person's life that illness causes. However, the naturalistic approach and modern science have brought enormous progress to medicine. Therefore what I propose is to augment, not replace, this approach.

I found phenomenology – the description of lived experience – to be the most helpful approach to augmenting the naturalistic account of illness. Phenomenology privileges the first-person experience, thus challenging the medical world's objective, third-person account of disease. The importance phenomenology places on a person's own experience, on the thoroughly human environment of everyday life, presents a novel view of illness. On the phenomenological account, illness is no longer seen merely as biological dysfunction to be corrected by medical experts. Because of phenomenology's focus on the subjective experience of the ill person, it sees illness as a way of living, experiencing the world and interacting with other people. Instead of viewing illness as a local disruption of a particular function, phenomenology turns to the lived experience of this dysfunction. It attends to the global disruption of the habits, capacities and actions of the ill person.

An example may help illustrate the shortcomings of naturalism. If someone suffers from depression, a physiological description of their illness will tell us very little, if anything, about the illness itself. Such a description may provide some information about brain function, neurotransmitters, serotonin levels and so on. But in order to understand fully what depression is, we must turn to the experience of depression: the loss of appetite, the dark thoughts, the listlessness and sense of doom and so on. If you tried to give a description of depression without recourse to any subjective experiences you would struggle to do so. This demonstrates that a purely physiological description of an illness is insufficient.

Another example is multiple sclerosis (MS). The physiological description would report localized failure of the central nervous system due to demyelination. The phenomenology of the illness, on the other hand, would tell of fatigue, the frightful experience of losing vision and other life-altering symptoms. (A fascinating phenomenological account of MS can be found in the work of S. Kay Toombs.) It would describe the difficulty of the everyday life of a person with MS, her fears of the future, her growing dependency on others and so on. Phenomenology does not deny the importance of the physiological description or of the clinical interventions offered by current mainstream medicine. It does propose to augment this approach to illness by emphasizing the importance of the first-person experience.

We are all ill at some point. The vast majority of us will die from some kind of illness. Everyone's life is touched by it to some extent. Illness and decay are universal features of life, human and non-human. So why is illness, as a woman with bowel cancer wrote to me, "a dirty little secret" that sick people share? What are the contents of this secret? What is the experience of being ill like? In order to begin to answer these questions we need a philosophical approach that enables us to both describe and articulate the experience of illness. This book unpacks the "dirty little secret" of illness in an attempt to make it less secretive and hopefully less lonely. The tension between the universality of illness and its intensely private and isolating nature is a riddle I hope to begin unravelling in this book.

WHAT IS PHENOMENOLOGY?

Phenomenology is a philosophical approach advocating a description of lived experience and consciousness. It focuses on what it is like to exist as humans in a world. It is a descriptive approach that rejects complex philosophical constructions of reality and puts aside questions about the nature of this reality. Instead, it focuses on the experiences of an individual, the ways in which we perceive things (*phenomena*) *as they appear to us*. Reality may be radically different from how we perceive it, but it is impossible for us to know how things "really" are. This is because everything we encounter, we encounter through our perceptual apparatus. Everything we experience, we experience subjectively. It is impossible for us to leap outside ourselves and have direct contact with the external world. It is also impossible for us to compare our experiences with how things really are, because any such attempt would only compare one experience with another. We cannot experience anything outside our experience. Therefore, phenomenology focuses on lived experience or on things as they appear to us (rather than how they are in themselves). This can be contrasted with a scientific, or objective, description of the world. Rather than trying to ascertain the true nature of objective reality (an impossible task to many phenomenologists), phenomenology suggests focusing on what is available to us, namely, the different acts of consciousness (such as thinking and believing) and our experiences and perceptions (things as they appear to us). As such, it is aptly named phenomenology: the *logos*, or science of phenomena.

Why is phenomenology a good approach for studying illness? Philosophers have been discussing the concept of illness for many decades. Over the past thirty years the debate has centred on two approaches to illness. One is the naturalistic approach, which is the prevailing one in the medical world. Proponents of this approach see illness (or disease, their preferred term) as biological dysfunction. They view illness in purely naturalistic terms, that is, only using concepts and entities that belong to the natural world. So having the flu, for example, would mean having a fever, an inflamed throat and a runny nose. These can be measured objectively and accounted for in third-person terms. The doctor can take the flu-stricken person's temperature, look at their throat and see that it is red and observe the runny nose. These facts can be seen by any observer and are therefore easy to capture in naturalistic, or objective, terms.

But is this a satisfying account of illness? Of course not, because the ill person also *feels* awful. She may have a headache or the shivers; she may feel cold or nauseous and so on. How are these things to be measured in objective terms? As Rachel Cooper writes, "no biological account of disease can be provided because this class of conditions is by its nature anthropocentric and corresponds to no natural class of conditions in the world" (2002: 271). This is where the phenomenological approach can complement the naturalistic approach. We shall see in a moment how phenomenology can do that.

The second approach is the normativist approach to illness. Normativism means using common social terms (or norms) to capture a particular phenomenon, in this case, the phenomenon of illness. Normativists think that the concept of disease is value-laden and that the naturalists are wrong in claiming that illness is biological dysfunction. In order to understand illness, say the normativists, we must focus on the way society perceives the ill person. For example, we believe that having flu is a bad thing. We feel sorry for the ill person and try to help her; we offer sick leave to enable her to rest. The normativist approach sees illness as something that must be socially evaluated as negative and not just a physiological process. But again, the first-person perspective is missing. Again the voice of the ill person herself is not heard.

A normativist account of illness would look to the social conception of the condition and the ways in which an illness may socially handicap the ill person. An example may clarify the difference between normativism and naturalism on this point. For some Indian tribes in South America, having dyschromic spirochaetosis, a skin disease, is a good thing. The resulting pattern on the skin is considered attractive and therefore the condition is desirable to these people. On the normativists' view, the skin condition is not an illness for the tribe because it is not perceived as negative in that society. For the naturalist, the skin condition is a disease because it detracts from the skin's function to protect the person from UV radiation and other external sources of harm.

These two approaches both have merits and have spawned a large literature. But there is a different set of issues pertaining to illness that is not captured by either approach. In this book I focus on what is left out of these two accounts, namely, the phenomenology of illness. That is, the experience of being ill: illness as it is lived by the ill person; the set of experiences – physical, psychological and social – and the changes that characterize illness. I view illness as a life-transforming process, in which there is plenty of bad but also, surprisingly, some good. The richness and diversity of experiences, the surprising and uncontrollable changes, the ways in which life is transformed by illness, are what this book is about.

The personal experiences recounted here are not just illustrations of philosophical ideas. Philosophy, too, is sometimes too objective, too distant from life, too focused on the third-person perspective. By combining the first- and third-person points of view, the subjective and the objective, the personal and the philosophical, I hope that the form of the book will reflect its contents. It is a genuine attempt to demonstrate the importance of having both perspectives present together. As such, this book is neither a personal story nor a purely philosophical reflection on illness. It is both.

OVERVIEW OF THE BOOK

Each chapter takes an aspect of life affected by illness and discusses it from a phenomenological perspective. By the end of the book, I hope to have created a collage of illness presented from the perspective of the ill person herself, from the first-person perspective, as opposed to the third-person perspective of naturalistic and normativist approaches to illness.

I begin by discussing the body and its transformation in illness through a phenomenological perspective. The French philosopher Maurice Merleau-Ponty (1908–61) proposed a fascinating analysis of the body and its relation to personhood. According to Merleau-Ponty, human existence is embodied and defined by perceptual experience. A change to the body and in physical and perceptual possibility transforms subjectivity itself. This view regards consciousness as embodied and shows that the human being cannot be understood without seeing it both as having a body and as having a world. The human being is by definition embodied and enworlded, so trying to provide an account of a human being that lacks these elements will result in a deficient account.

On this view, the body is not an automaton operated by the person but the embodied person herself. We *are* our bodies; consciousness is not separate from the body. Disease, therefore, can no longer be understood as a mere physiological process that affects the person only secondarily. This is not just the trivial view that our lives and subjective experiences are affected by disease, but a deeper conceptual shift. On the phenomenological view, disease cannot be taken as a mere biological dysfunction, because there is nothing in human existence that is purely biological. We *are* embodied consciousness, so consciousness is inseparable, both conceptually and empirically, from the body. Therefore the concept of illness must be reconceived to take this unity into account.

Next, I look at changes to the world of the ill person. When we think about the term "world", two meanings are available: we may be referring to the physical world or to the sociocultural world. Changes to the way in which the ill person experiences her physical world are obvious. The topography of an area may be perceived differently by a person in a wheelchair than by an able-bodied person. Similarly, distance cannot be conceived as objective. What may seem near and easy to the healthy could be distant and difficult for the ill. The ill person may need to reconceive her physical world and make changes to it as a result of the changes in her bodily abilities. Distances increase, hills become mountains and stairs become obstacles rather than passageways. The physical world is altered for the ill person. Chapter 1 offers a geography of illness, showing how the surrounding world and the interaction with it change in illness.

The social world is also significantly altered by changes to the ill person's abilities. This change is captured by phenomenology's attention to the relationship between agency and the body. The possibility of agency, the ability to act effectively in the world, is inherently linked to the ability to assert oneself, perform actions and carry out activities that promote one's goals. But when one is ill, one's ability to perform actions becomes limited. In other words, the fundamental role of the body as enabling agency requires special attention in illness.

Changes to the physical body will have not just a physical effect, but may curtail the ill person's actions, her ability to carry out goals and operate effectively in the social and cultural world. For example, wheelchair-users often complain of being ignored, being talked down to or being treated as unable to understand what is said to them. A relatively small change in posture (sitting rather than standing) has substantial consequences for their social status, the way they are treated. It is not only their body that is limited by their inability to walk; their agency, too, is transformed by the physical limitation. The embodied nature of agency and the modification of agency by bodily limitations is therefore a central aspect of illness and, again, one that has been overlooked by the naturalistic approach.

In Chapter 2 I discuss the social world and its transformation in illness. I ask how the ill person is seen by different social agents in various situations. The encounter with healthcare professionals will come under focus, as will meeting strangers. I then turn to friendship and the strains placed on it by illness. Betrayal and disappointment, how illness poses a threat to intimacy and fear of the diseased body are explored.

Another way of thinking about illness is to see it as a state of inability. When ill, one is unable to do some things, perform particular roles and engage in certain activities. In Chapter 3 I explore the notion of illness as dis-ability: as being unable to do, to act and to move freely. I examine this notion in relation to the German philosopher Martin Heidegger (1889–1976) and his definition of human existence as "being able to be". Heidegger's definition of the human being is best understood by his notion of projection. Projection means throwing oneself into a project, through which the human being's identity is defined. For example, if my project is being a teacher, I project myself accordingly by training to be a teacher, applying for teaching positions and so on. This, Heidegger claims, is the essence of human existence: the ability to *be* this or another thing, to purposefully project ourselves into the future.

But in some illnesses, especially mental and chronic illness, a person's ability to be, to exist, is sometimes radically curtailed. Certain projects must be abandoned and sometimes, as in severe psychosis, the possibility of having a project at all becomes difficult. I present this problem and then propose a new view of Heidegger's notion of existence as the ability to be. For example, what if we allow radically differing abilities to count as forms of human existence? How flexible are human beings? How much can we adjust our projects and plans in the face of ill health? I explore possible answers to these questions.

But there is a more positive view of illness. Such a view is expressed in the idea of health within illness. Examining recent nursing literature informed by a phenomenological approach, I develop the notion of health within illness and discuss the results of studies looking at reactions ill and disabled people have to their illness and how these affect their well-being. Rather than measuring the experience of the ill person in objective parameters – how far from the norm she is – health within illness focuses on experiences of personal growth, adaptation and rediscovery. A phenomenological approach enables the expression of these experiences in order to give a more complete description of the altered relationship of the ill person to her world and a better understanding of her experience.

Here I also provide a positive answer to the question: can seriously ill or disabled people have a good life? I develop the idea of illness as an extreme case of lived experience, one in which the usual "norms" are rewritten and require significant adjustment and creativity. Two central ideas – that illness induces adaptation and that adversity is the source of creative responses to it – serve as the basis for this positive reply.

The fear of death is part of any serious illness. Having a poor prognosis at any age raises questions such as: should I fear death? How can I prepare for my own death? Can a person with a terminal-stage illness have a meaningful life despite her knowledge of her imminent demise? Should ill people think about death, try to come to terms with it, or should they try to ignore the inevitable? In Chapter 4 I pit Heidegger's and Epicurus' contrasting attitudes to death against each other.

On Heidegger's view, mortality is the defining feature of human existence. In order to make sense of our life, we must understand it as finite. His notion of "being towards death" is used to describe human existence as death-bound and temporally finite. As a phenomenologist, Heidegger faces a certain paradox in his discussion of death: death is not something we experience. Therefore he proposes not a phenomenology of death, but a phenomenology of attitudes towards death. "Being towards death" is a term that captures this specific angle that Heidegger wishes to analyse.

From a very different approach Epicurus presents us with rational arguments designed to combat our fear of death. If death is the state of non-existence, there is nothing to fear in death. There is no point worrying about it while we are alive. Are life and death mutually exclusive, as Epicurus suggests, or are they intimately intertwined, as Heidegger thinks? These issues are discussed in Chapter 4, in which I examine the fear of death as an integral part of illness.

In Chapter 5, the final chapter, I introduce the notion of philosophical therapy. Continuing the discussion of Epicurus, I turn to his idea that philosophical arguments should be "medicine for the soul". For ancient Greek philosophers, philosophy was not an academic subject to be studied in isolation from everyday life. On the contrary, philosophy was a set of practical skills intended to enhance and improve life. The idea that philosophical reflection can be an aid to challenging times will be applied to illness. In the same way that Epicurus offered arguments against the fear of death, we shall see how a certain view of time, of living in the present, may be more amenable to living with illness. By using philosophy to address a practical question, namely "How can I live well with illness?", Chapter 5 brings together the first-person perspective of the lived experience of illness and the third-person perspective of philosophical reflection.

In the hope that some healthcare practitioners will be reading the book, I make some suggestions along the way about how a phenomenological approach could inform their important work. There are some practical suggestions on how a phenomenological approach could be incorporated into current healthcare practices and how the patient's point of view could be made more relevant and informative.

In what follows, I share personal experiences and insights in the hope that these will illuminate the philosophical ideas discussed here as well as demonstrate the importance of the first-person perspective. I describe real encounters but names and details have been changed to avoid identification.

When I speak of illness in this book, I refer to physical illness. Since I have no experience of mental illness, I thought it prudent to limit myself to physical illness. This is not to deny the importance of mental illness or the suffering experienced by those who are affected by it. I also make few references to disability (unless speaking about my own disability). Again, this is not to deny the significance of disability, but to avoid speaking on behalf of people who may be affected by different matters. I presume that there is some overlap between the experiences characteristic of the three conditions and that my analysis may be relevant to mental illness and to disability, but I make no claims about either.

I am not claiming for a moment that my experiences are universal or even similar to those of others. I share them in the hope of making illness a little less scary, less anonymous, by talking about what happened to me. It is my hope that illness, a near universal event affecting almost everyone, will be illuminated anew upon viewing it from a phenomenological stance.

ONE
The body in illness

It is Christmas 2004, the height of summer in the southern hemisphere, and I am in New Zealand. My friends and I are on holiday, touring the South Island for two weeks. Part of our trip is a three-day coastal walk in Kaikoura. We are walking towards the farm we are staying in on the eastern coast of the South Island. On the way we see dolphins and seals and wood pigeons. The air is fresh and the scenery beautiful. I am eager to hit the trail and confident because I have been exercising a lot lately and feel fit and full of life. I have become a health freak, eating little fat, spending forty-five minutes a day on the Stairmaster and lifting weights. I bounce and walk forwards, happy, energetic, bursting with joy. We walk at a brisk pace, chatting and enjoying the views and the sunshine.

The terrain changes and we are now walking uphill. Suddenly, things become difficult for me. I lag behind; I can no longer chat to my friends. I stop and pour out the water I am carrying. Perhaps I am carrying too much weight? I try to walk on, but something is slowing me down. I have to stop frequently to catch my breath. No matter how slowly I walk, I still have to stop. The trail that seemed so inviting and beautiful is now harsh and endless. Eventually, I lag almost an hour behind the group. My ever patient sister-in-law, Mona, notices I am struggling and slows down to my pace. She stops with me, pretending she wants to look at the views. I am worried: how could I be so unfit? Why isn't my body responding to all the exercise? I thought I'd be leading the group, but instead I am soon labelled the slow one, the struggler.

In the months that follow my husband and I invent a string of explanations for my breathlessness. Maybe my lungs are small? Perhaps I have asthma? Maybe I have a tiny chest infection? I return to the gym with greater ferocity and determination than before and sign up to an additional kick-boxing class. I don't go to the doctor, and will not do so until two years later, when my breathlessness has become so prominent and abnormal that these feeble excuses no longer seem reasonable. But the sense of uncertainty, the struggling, the inability to understand my own body's responses, have been constant companions to me since.

The betrayal of the body, and the increasing alienation from it that an ill person experiences, is the main focus of this chapter. So how is the experience of an ill body different from that of a healthy one? The phenomenological approach of Merleau-Ponty provides a fascinating account of this difference. It is Merleau-Ponty's emphasis on perception and on the centrality of the body to human existence that I find particularly illuminating in relation to illness.

Merleau-Ponty sees the body and perception as the seat of personhood, or subjectivity. At root, a human being is a perceiving and experiencing organism, intimately inhabiting and immediately responding to her environment. To think of a human being is to think of a perceiving, feeling and thinking animal, rooted within a meaningful context and interacting with things and people within its surrounding. By taking this approach, Merleau-Ponty responds to a previous, intellectualist (as he calls it) definition of the human being provided by the seventeenth-century French philosopher René Descartes (1596–1650). Descartes defined us as thinking, abstract souls who temporarily and contingently occupy a physical body. Descartes's approach is known as "dualism" because it postulates two different substances: spatial or extending substances such as physical objects, and thinking substances such as minds.

Merleau-Ponty's aim was to correct this erroneous view and, while avoiding the materialist reduction of mind to matter, to emphasize the inseparability of mind and body, of thinking and perceiving. His approach can be thought of as holistic with respect to the human being. We cannot divide a person into a mental and a physical part, because the two are *de facto* inseparable. Any mental activity must have some physical action underlying it (for example, some neuron firing in the brain). It is impossible, on Merleau-Ponty's view, to think of a purely mental action because mental activity, abstract as it may be, is always embodied. Additionally, for us to acquire abstract notions and concepts requires experience of the world. So, for example, our concept of the colour red arises from seeing red objects. The concepts arise from sensual, perceptual experience. If we take these two arguments together, we can see the grounds for Merleau-Ponty's claim that there is no mind that is independent of body in the strict sense.

Similarly, physical action cannot be seen as mechanical manipulation mysteriously governed from a distance by mental commands. The body is not a passive vehicle simply awaiting instructions from the mind. Nor is it a system of pulleys and levers (as seventeenth-century mechanistic philosophers thought) that only comes to life when infused with a soul. Rather, it is an active entity, capable of goal-oriented action and intelligent response to the environment. The separation between mind and body does not make sense. Moreover, the strict separation between an internal realm and an external world does not make sense when we think about how we actually experience our body and the world, as a seamless unity.

Instead of artificially separating mind and body, Merleau-Ponty emphasized the centrality of the body and gave an account of how the subject inhabits it. This more organic and biological view of the human being as a human animal (which also has culture, sociality and a meaning-endowed world) sees the body as the seat and *sine qua non* of human existence. To be is to have a body that constantly perceives the world through sight, touch, smell and so on. As such, the body is situated and intends towards objects in its environment. Human existence takes place within the horizons opened up by perception.

Thus for Merleau-Ponty the body is a body-subject, engaged in a "primordial dialogue" with the world. This dialogue is pre-reflective, absorbed engagement with the environment, which can be easily understood by thinking about everyday activities. For example, going for a walk is such a dialogue of the body with the environment: the legs propel the body forwards, the labyrinth in our ears keep us upright and balanced, the eyes provide visual information about the path ahead and any obstacles to be negotiated, and so on. This kind of dialogue with the environment requires the constant taking in of information and constant recalculation of route, speed and muscular effort.

A second dialogue takes place between different body parts and types of information. This synthetic activity unifies the information coming from the eyes, legs, muscles and so on to create a unified experience of walking. The whole time this complex interaction takes place, the walker could be avidly discussing Nietzsche, paying no conscious attention to her body. This does not make her disembodied, and does not bring back Cartesian dualism. It simply shows that embodiment is a background condition for subjectivity. This holds true even if no attention is paid to the body. Whether playing tennis or working out a mathematical problem, both activities, and the whole spectrum in between, are possible only in virtue of having a body, existing as embodied in a world.

Many of our actions, particularly everyday routine actions, are pre-reflective: they are the product of habit rather than conscious reflection. A complex web of such habits makes up our behaviour. Our habits and ordinary ways of engaging with our environment create a familiar and meaningful world. Against this often implicit background activity, reflection and conscious thought take place. Normally we pay attention to what is consciously preoccupying us at a given moment, for example, thinking about a philosophical problem, rather than about the cup of tea we are preparing while thinking. But Merleau-Ponty wants to focus on the significance and sophistication of this background and moreover to understand how it enables conscious thought to take place on top of it, as it were.

The body is the centre of his investigation. It is a unique kind of object for Merleau-Ponty. The body is, of course, a physical thing, an object that can be weighed, measured and described using purely physical or naturalistic terms. But it is also the source of subjective feelings, perceptions and sensations, the seat of subjectivity, and consciousness. As such the body is a subject-object, a unique being that can be experienced both from a third-person point of view (we see other people, measure their height, observe their eye colour) and from a first-person point of view (I feel myself sitting on the chair, I am thirsty, I stretch my arms and experience my muscles distending and releasing).

Merleau-Ponty uses the simple example (from Edmund Husserl's *Cartesian Meditations*) of two hands touching each other. Each hand is both touching, active, sensing the other hand, and also being touched, passive, being sensed by the other hand. It is this view of the body as being both an active touching subject and a passive touched object that reunites the mind and body, the first- and third-person points of view, and expresses most clearly the unique position of the body.

Merleau-Ponty develops the notion of bodily intentionality. Intentionality was originally conceived by Franz Brentano and Husserl as a relationship between mental phenomena and their objects. It is the relationship of being *about* something, or intending *towards* something. For example, if I wish to eat ice cream, ice cream is the intentional object of my desire.

Interestingly, only mental phenomena are intentional, or *about* something. Wishing for ice cream, or believing that my bicycle is in the shed, are examples of intentionality towards ice cream and bicycles. Every mental act such as believing, desiring and so on, must be about something, or in other words must have an object. Thoughts are *about* something, beliefs and desires are *about* something, but physical objects cannot be about anything. Physical objects are not mental phenomena and therefore lack intentionality. A shoe cannot, in principle, be about anything. This feature of "aboutness" is often regarded as the defining mark of the mental.

Merleau-Ponty took on board the idea of intentionality but refused to accept that only mental phenomena can have this property of intentionality or "aboutness". He extended the notion of intentionality to include bodily intentionality. This is the body's intending towards objects, directing itself at goals, and acting in a way that is "about" various aims and objects. For example, if I reach out to pick up a cup of tea, my hand intends towards the intentional object, the cup. The position of the hand, the direction of the movement, the arching of the fingers are all directed at, or intended towards, that cup.

An intentional arc, as Merleau-Ponty calls it, connects my body to the cup of tea. This intentional arc makes sense of a collection of disparate bodily movements, unifying them into a meaningful action: intending towards the cup of tea. In this sense we could say that bodily intentionality is analogous to mental intentionality. Some philosophers make the stronger claim that bodily intentionality is primary to and the foundation of mental intentionality. They claim that there can be no mental intentionality without bodily orientation in a world: that mental intentionality is always underpinned by bodily intentionality.

What are the implications of this bodily intentionality? This notion contributes to our view of the body as an intelligent, planning and goal-oriented entity. The body is not a passive material structure waiting for mental commands, but rather is actively engaged in meaningful and intelligent interaction with the environment. The body *knows*, so to speak, how to do many things, how to perform minute and complex actions, how to achieve a plethora of goals from ice-skating to driving a car. Through its directedness the body executes actions that are not merely random physical movements, but intentional, planned, goal-directed movements. Moreover, the movements only have meaning when understood as aimed at a goal. "For us the body is much more than an instrument or a means; it is our expression in the world, the visible form of our intentions" (Merleau-Ponty 1964: 5).

The body responds to the environment in an ongoing dialogue. Everything else depends on the body's ability to perform, predict and react appropriately to stimuli. Thus the body is the core of our existence and the basis for any interaction with the world. "The body is our general medium for having a world" (1962: 146). All our actions and goals have to be rethought in light of this new role accorded to the body or, more accurately, new recognition of the role that it was playing all along.

THE ILL BODY

Having seen how central the body is to any notion of agency or subjectivity and to achieving any goal, we can now ask what happens when the body loses some of its capacities and becomes unable to engage freely with its environment. In illness, and more pointedly in some cases of chronic illness and disability, we find a need to rethink the body's ability to engage with the world, its ability to provide movement, freedom and creativity as it did before.

So how should we think about illness? If we go back to Merleau-Ponty's view of the body as both object and subject, the *ambiguity* of the body, as he calls it, we can see an important dimension of bodily experience exposed in illness. This is the difference between the biological and the lived body. The biological body is the physical or material body, the body as object. The lived body is the first-person experience of the biological body. It is the body as lived by the person. Normally, in the smooth everyday experience of a healthy body, the two bodies are aligned, harmonious. There is agreement between the objective state of the biological body and the subjective experience of it.

In other words, the healthy body is transparent, taken for granted. We do not stop to consider any of its functions and processes because as long as everything is going smoothly, these are part of the bodily background that enable more interesting things to take place. So while digestion, fluid balance and muscular performance are going well, we do not experience them consciously. They silently and invisibly enable us to compose symphonies, have coffee with friends and daydream while walking the dog.

It is only when something goes wrong with the body that we begin to notice it. Our attention is drawn to the malfunctioning body part and suddenly it becomes the focus of our attention, rather than the invisible background for our activities. The harmony between the biological and the lived body is disrupted and the difference between the two becomes noticeable.

We can also think about the body using the analogy of an instrument or a tool. Take a pen, for example. We normally use a pen to perform a task, say, to write a letter. While using the pen, we do not notice it. It is inconspicuous, a means to an end. Our attention is focused on the end – writing the letter – while the means are relegated to the background. But when the pen fails to write, the car refuses to start, the milk bottle is empty, they suddenly become the centre of attention. They cease to be an invisible background enabling some project and become stubborn saboteurs.

This inconspicuousness characterizes tools and even more so for our body. Whereas we can throw out the useless pen and grab another, our body stands in a very different relation to us. Our body cannot be replaced, thrown out or even repaired as readily as we would like it to be. My head with a headache remains attached to me, and becomes increasingly conspicuous, increasingly disabling. It is precisely *because* our body is not a tool that its dysfunction is so intimately linked to our well-being. Whereas my malfunctioning car can be sold and a new one bought, my body *is* me. This is an essential feature of our embodied existence that is brought out by illness. Illness is an abrupt, violent way of revealing the intimately bodily nature of our being.

Another reason the difference between the biological and the lived body emerges in illness is that the lived body is in large part habitual. It is used to performing certain tasks at a certain speed, in a certain way. Think of the way you do something that you do routinely: shave, play tennis, chop vegetables, sew, play the piano. These actions can be performed expertly, efficiently, smoothly, because they have become habitual. Our body learned to perform them and with each repetition the habit is reinforced, incorporated further into our bodily repertoire. We may perform some actions with little or no attention. Again, the actions are harnessed to the goal of the activity: getting to work, a cooked meal, an ironed shirt. It is only when we watch a novice, say, a child learning to ride a bicycle, that we appreciate the difficulty of the activity and the level of expertise our bodies have acquired.

Our everyday activities rely on the interplay between the body as it is experienced now and the habitual body. The habitual body provides the framework, the expertise, while the body as it is experienced now provides instant feedback from the environment, different sensations, feelings of satisfaction and so on. Our expert bodies are the product of many years of habituation and practice. The practice can be conscious and structured, as in taking dance lessons or driving instruction, or unstructured and spontaneous, as in children's play.

This acquired expertise, the effortless ease with which we perform habitual tasks, relies on two things. One is the continuity of the experiencing subject. The same person who learned to play the piano as a child is now playing Mozart's Fantasy in D minor. Memory and continuity are essential for maintaining personal identity over time. The second is that the actual capacities of the biological body remain intact. These physical capacities underlie any attempt to rehearse or master new skills. A common example would be that of an adult attempting some childhood feat, such as a somersault or back dive that they used to perform as children. If she is out of shape, if her muscles and flexibility are no longer as good as they used to be, the habitual body will encounter the resistance of the biological body. The habit may still be there, but the biological body has changed and is no longer cooperating with the ease and agility it once had.

Another example given by Merleau-Ponty is the phantom limb. A phantom limb is the sensation, which may be painful, emanating from a body part (usually a limb) that has been amputated. The phantom limb feels painful or itchy, but the real limb has been removed, sometimes a long time ago. This phenomenon has been a riddle to physicians for centuries.

How can a limb that is no longer there cause any sensation at all? What exactly is hurting when the phantom limb hurts? This baffling phenomenon could not be accounted for by a physical naturalistic explanation, because the nerve endings that seem to be sending pain signals to the person's brain are simply not there. How can you experience pain or tickling in a hand that was amputated years ago?

If we return to Merleau-Ponty's distinction between the biological and the lived body, we can explain the phantom limb as a rift between the biological body and the lived experience of it. The biological body has no limb, but the lived body feels that limb as present, painful, itching. It is no use telling the person that their limb is not there, or denying their lived experience. The phantom limb is the expression of the body as it used to be, based on decades of having a body schema with four limbs. This habitual body is not destroyed instantaneously when the biological body loses a limb. The body schema must be reconstructed, new bodily habits and movements must be created, to compensate for the loss. And sometimes, as in phantom limb cases, the lost limb continues to exist, so to speak, in the lived experience of the person.

Another example of the rift between the biological and the lived body is anorexia nervosa. If we look objectively at the biological body, we may see a skeletal, emaciated body. This is the biological body, whose thinness can be measured by weighing it or calculating its body mass index. But if we ask the anorexic to describe her body, she may say that she experiences it as obese, cumbersome, large. The subjective body, or the body as lived, is a fat, monstrously corpulent body. As we know, denying this experience by making an appeal to objective facts is unhelpful. Here, again, we can see the rift between the body as it is objectively and the body as it is experienced.

Understanding this rift gives us the tools to describe the impact of illness. Because the body under Merleau-Ponty's description plays a central role, and because illness changes the body, the impact of illness is more significant than we may have previously thought. We can now begin to see how being ill is not just an objective constraint imposed on a biological body part, but a systematic shift in the way the body experiences, acts and reacts as a whole. The change in illness is not local but global, not external but strikes at the heart of subjectivity. Because "I do not bring together one by one the parts of my body; this translation and this unification are performed once and for all within me; they are my body itself … I am not in front of my body, I am in it, or rather *I am it*" (Merleau-Ponty 1962: 150, emphasis added).

MY ILL BODY

We are creatures of habit. I have already forgotten what it was like before. Before I was ill. Before I lived in the shadow of chronic breathlessness, fear of infection, and the endless need to take care of things. The endless need to arrange oxygen deliveries, attend scans, renew my prescription, pick up drugs, take drugs, attend consultations – in short, the need to manage my illness.

My illness has become part of my life at an incredible pace. At first it was an external disaster; something that wasn't meant to happen to me; something extraordinary, while I was ordinary. In the first week after my diagnosis I would wake up, blissfully ignorant of the new reality of my life. After a moment, as I awoke, the realization that something had gone horribly wrong would crash down on my sleepy head. These moments of being awake but not remembering I was ill quickly disappeared. A new era began. My illness was internalized and became part of my life, part of me, my body.

My body adapted with astonishing alacrity to new limitations. I quickly forgot how things were before. Within a year my physical habits were entirely different. Whereas in the first months my body would attempt a brisk pace, hurrying up stairs, physical impatience, these movements have been erased from my bodily repertoire. While my memory still contained images of mountain-top views and the inside of a gym, I could no longer remember what it *felt like* to run, to work out, the euphoric sensation of healthy exertion, the effortlessness of being young and healthy. New habits were formed and a new way of negotiating the world was incorporated into my physicality. Blissful forgetfulness of the pleasures of physical movement accompanied them.

In my pre-illness days I made plans and wanted the usual goods life offers us. I thought my wishes were mediocre, underpinned by a modest, implicit list of expectations: to be healthy, to be happy, to be safe. After I became ill, these wishes began to seem exuberant. Did I really expect all that? Did I really think that this list of infinitely complex, luck-dependent elements would just come true as a matter of course?

I began to think that I was asking for too much, that we all routinely ask for too much. That nothing could go smoothly all the time, let alone everything all the time. But I had another card up my sleeve. I was good. I ate a healthy diet. I exercised. I didn't smoke or drink. I took care of myself. When I compared myself with friends, many of whom have been heavy smokers for over twenty years, I felt I deserved to have the lives they have, just by chance, more than they do. I fell into the beginner's trap of suffering and asked: why did this happen to me? This question had no answer, or at least no answer we know of with current medical knowledge. It was an arbitrary stroke of very bad luck.

My expectations had to change, and change fast. My wish list narrowed down to having one item only: I want to live. I don't expect the shadow ever to lift from my life and disappear. I don't expect to have a family, to be able to pack up and go on a trip, to celebrate my sixtieth birthday or to spontaneously book a last-minute flight to Greece. I don't expect to feel better. My only wish is to hang on to what I have now: being alive, having a moderate quality of life, doing some of the things I love. That's plenty, I tell myself. And it is this veneer of normality, this slightly sad and inappropriate sense of luck, that sustains me in my illness.

It is when I encounter pity, when people show some of their incredible sadness and anxiety for me, when people respond with shock and horror to my condition, that my story crumbles. And then I know: things work out for most people most of the time. Only 160 women in the UK and a few thousand worldwide have been diagnosed with LAM. It is true that other people have different problems. But if I look at my friends and acquaintances, most of them are entirely healthy. Most of them did go on to have healthy babies. Most of them have all the things that were on my all too human wish list, but take them for granted. Most of them live in hubris of which they are blissfully ignorant. Instead, they have the naive, unreflective sense of desert. Why shouldn't things work out for me, they ask?

In the early days after my diagnosis I couldn't think at all. I didn't dare read about my illness or learn anything about it. I felt that any more information would only bring with it bad news, more horror, additional grim facts to petrify me. I suffered from what Joan Didion calls "magical thinking": the irrational, self-blaming, mystic thought that is apparently common in situations of distress. I blamed myself for writing a book on death. I blamed myself for going to the doctor so late. I blamed myself for being arrogant and not budgeting for something like this from the beginning. I blamed myself for daring to have a wish list.

Later, as I adjusted to my situation, I felt increasingly angry. I spent several months asking: why did this happen to me? I felt sorry for myself. I cried for days, with grief for the children I thought I would never have, for the short and crippled life available to me. When I walked through the park I would look at the young mothers playing happily with their children and a wave of envy would wash over me. You will never have this, I would tell myself over and over: the sense of security, the naive belief in the goodness of life, long lazy afternoons in the park, mundane concerns about redecorating and scraped knees.

It was only through long conversations with another woman with LAM that I stopped the vicious circle of envy, self-pity and despair. Grazyna, who has been ill for a long time and suffered markedly worse ill health than me, became a dear friend. I felt I could talk to her because she, too, had LAM and she, too, faced the same challenges as me. She saved me from myself by listening with great compassion but also by refusing to accept the destructive views I voiced about our condition. She told me to stop moaning, stop making a fuss about trivial things and start appreciating what I still had. When I complained about being stared at on the street because of my oxygen cylinder, she told me to look away from myself. To stop thinking I am so interesting to others and, instead, enjoy the freedom the oxygen provided.

Her refusal to accept my self-pity worked. I realized I needed discipline. I needed discipline in my illness management and I needed emotional discipline. I needed to take the drugs and see the doctor and have the scans and do the breathings tests and order the oxygen and renew my prescriptions and sleep with the oxygen mask even though it caused my nose to bleed. I needed to walk as much as possible and go to yoga three times a week and eat a healthy diet and take the vitamin supplements and get a flu jab every autumn and patiently tolerate invasive tests and painful medical procedures. This was the easy part.

Developing emotional discipline was much harder. I needed to learn to say no to negative feelings. I needed to learn to go and sit in a cafe when writing at home became morbid. I needed to stop looking at other people's lives and making up stories about their happiness. I needed to open myself to the knowledge that other people suffer too, that there are other kinds of pain and sources of misery. I needed to recite, even by rote, all the good things in my life and to cultivate that perversely optimistic feeling I had deep down inside, that everything was going to be OK.

I needed to learn to stop caring about what other people have. To focus on good things and look away from suffering. To avoid sad films and novels. To stop noticing children in the park. I had to kill a part of me to achieve that, but the trade-off was essential. I had to stop caring about people's stares and sometimes rude comments and learn to be rude back. I had to develop blindness to certain kinds of looks, and deafness to certain kinds of overtures. I had to learn to stop being nice at my own expense. I had to learn to say "I am disabled" when booking hotel rooms. I had to force myself to walk into a room full of people with my oxygen on and my head held high, watch people's eyes widen with surprise and then look away, embarrassed.

I needed to immerse myself in new projects. And so we bought a house and adopted a rescued lurcher, Laika, named after the Soviet dog sent to space in 1957. I spent a long time training and walking her and my fitness duly improved. At first, the snail's pace crawl up the hill to our local park was utterly painful. But as time passed, I became a little fitter, or a little less conscious of how slow I was, and the walks became truly pleasurable. I discovered I enjoyed gardening and spent many afternoons in the sun, bending and lifting, watering and snipping away happily, at my limited pace. I went on walks, and cycled on my electric bike, the most ingenious gift I have ever received. And sometimes if it weren't for the baffled stares, I would almost forget anything was amiss. Things became abnormally normal.

The process of normalization is a curious one. It made me admire my body in its automatic, tacit wisdom. I learned what complex processes of compensation take place when lung tissue is impaired. How the body automatically increases the number of oxygen-binding red blood cells. How the heart works harder, pumping the blood ever faster through the lungs in an attempt to absorb more oxygen from a reduced surface area.

I began to self-censor actions and movements that caused me breathlessness. Every time I tried – and failed – to do something that was too strenuous, my body stoically registered the failure and thereafter avoided that action. The change was subtle, because this happened by stealth. The miraculous result created by my body's adaptive abilities was that I stopped feeling so acutely all the things I could not do. They were quietly removed from my bodily repertoire, in a way so subtle I hardly noticed it. The creeping changes added up, of course, but their impact was limited by the constant adjustments occurring at a subliminal level.

My habits changed. I began to plan each trip upstairs, compiling a small list in my head to avoid unnecessary journeys up one flight of stairs, a journey that would leave me gasping for air. I planned meeting places with friends according to gradient. Some parts of town or of buildings became off-limits to me: the top of Nine Tree Hill, my colleague's office on the fourth floor, top-floor flats. I began to feel amazed at people walking uphill while chatting.

My daily walk with Laika was planned meticulously to avoid unnecessary ascents. After being late to dozens of appointments I began to add fifteen minutes to my journey time. I began to arrive early at train stations and wait at the top of the stairs until the very last minute, in case of a platform alteration that would force me to climb up the stairs again. For many months I resisted the lift. I would walk up stairs slowly, excruciatingly breathless; people would often stop to ask whether I was OK. The day came when I gave up stairs and stopped trying to pretend I was the same as before.

I adjusted my walking pace to a slow, measured one. Whenever walking, part of my mind is now preoccupied with my breathing: do I need to slow down? Do I need to adjust the oxygen level? Should I try this hill? My days are now divided into "good days" and "bad days". Good days are exhilarating, elating. On a good day I can walk briskly (with oxygen) on the flat, walk up Old Ashley Hill at a very slow pace, get upstairs with only one stop on the way, hold yoga poses as long as the rest of the class. Bad days feel like death. On bad days my body closes in on me, reminding me constantly of my inability to do so many things. On a bad day, a trip down one flight of stairs to buy a cup of coffee is too much.

I remember still trying to do things. I remember a game of American football in Australia, on the lawns of the Australian National University, where I used to teach. My nimble friend Karen, who I was guarding, dashed ahead. I gave chase, but before I knew it she scored a touchdown. I remember playing basketball with friends and having to stop to rest every few minutes. I remember trying to run on the treadmill at the gym and having to stop. I remember climbing out of a valley in the Blue Mountains in Australia and being left far behind by the others. All this time I thought I was weak-willed, lazy. I thought I needed to try harder.

Eventually, my body learned to stop trying. It learned to give up its habits and form new ones. I was told about the dangers of secondary pulmonary hypertension, damaging the heart, which is working so hard to pump blood into the lungs. I was told about the danger of a collapsed lung and respiratory failure. When I sleep alone I take the phone upstairs with me. Just in case.

My body has adjusted, its habits now largely transformed by its newly acquired limitations. But my mind still aches in its desire for freedom, for joyful movement, for physical abandon. I often dream about running. I am shooting ahead, running at an incredible pace, my feet hardly touching the ground. I run and run, unlimited by my breathing, exhilarated by my self-generated speed. It is these dreams that remind me of what has been taken away from me: the bodily freedom I hardly ever thought about when it was mine.

TWO

The social world of illness

Empathy. If I had to pick the human emotion in greatest shortage, it would be empathy. And this is nowhere more evident than in illness. The pain, disability and fear are exacerbated by the apathy and disgust with which you are sometimes confronted when you are ill. There are many terrible things about illness; the lack of empathy hurts the most.

I am in the respiratory department for my breathing tests. I begin preparing several days before the test. I always brace myself for a decline, telling myself: you *know* it will be worse this time. A further deterioration brings with it a further shrinking of my world, fewer things I am able to do easily, or do at all. Every month, as my breathing deteriorates, I wonder what will go next. Will I have to give up my electric bike? Will we have to install a downstairs toilet? Will I be able to continue practising yoga? Seeing your capacities diminish, your world becoming smaller and harder to negotiate, is never easy. Most people experience decline over decades.

But seeing your abilities shrink at a terrifying pace at the age of thirty-five is horrific. Nonetheless, I know I must have the breathing tests.

A lovely physiologist, Simone, is usually there, chatting to me, telling me about her boyfriend. But this time she is not there. Another physiologist, sullen and unfriendly, leads me to the test room. She sets up the machine without saying a word, hands me a tube and tells me to blow into it as hard and as fast as I can. I take a big breath. I exhale into the tube. I blow hard. My face goes red, my body tenses, my shoulders quiver with the effort. I want good results. I want the same results as last time. I want to be stable, oh, how I want to be stable. I blow until I feel like fainting. I want to be able to blow the same meagre 1.4 litres of air out of my lungs as I did last time. (This test is called FEV1, the forced expiratory volume expelled in the first second of exhalation. A normal result for a woman my age is about 3 litres.) I blow. My lungs are empty and I feel dizzy from the lack of oxygen. But I keep blowing as hard as I can, stretching the beleaguered membranes of my lungs well beyond their capacity. The needle, unresponsive, barely shifts. It crawls upwards, marking the diminishing amount of air I exhale. I sit down, panting. I've done my utmost. I've blown myself away. But I know I failed. I know I declined.

I ask her for my result. "1.1," the nurse says, with no trace of emotion in her voice. I try hard not to cry, but panic and despair get the better of me. I choke on my tears. Crying involves a lot of breathing at the best of times; with respiratory illness, it is downright difficult. I sob quietly, bitterly, the way defeated people cry. I lament my helplessness, my body's betrayal. I can't do it. I can't breathe properly. I cannot breathe. All those hours at the gym, kick-boxing classes, strength training, runs – all to no avail. My illness is stronger than my body, stronger than my will. I've lost 300 precious millilitres of lung function over the past three months. The equivalent of what a normal adult would lose over a decade. I look at the physiologist. She stands there, stony but for her slight impatience. Now I'm crying and can't do the other tests. I'm spoiling her day, getting her behind schedule. I collect myself; ask her for a glass of water. A sulky hand presents me with a dripping paper cup. She doesn't look at me or say anything. I am alone.

I later reflect on the encounter with the physiologist. What sort of training has made her able to stand there, saying nothing, offering no word of comfort or distraction? Does she do this every day, to all her patients? Does she feel anything but annoyance towards me? Is this exchange sanctioned by the National Health Service? Does she think of me as a person? I can't ask her these questions. She probably won't even remember me. I know I failed the unwritten law of the medical world, where everything is impersonal, where news of deterioration and terminal illness are to be met with dry eyes and a steady gaze. And within this world, my human failure will be held against me, while her failure to be human does not even have a name. I never went alone to another breathing test. My infinitely patient husband now insists on coming to every single hospital appointment. He sits in the room, sipping tea and joking with the physiologist, another physiologist, a very young one who smiles and chatters, making the tests so much more bearable. I don't like having to have him there. The test takes over an hour and I am always worried that he is bored, thinking about the work he's missed, a class that will have to be rescheduled. He has enough to deal with. I don't want to involve him any more in the tedious mechanics of being ill. But I cannot face my own decline, my body's failure and its terrible implications, alone.

These kinds of encounters with health professionals have repeatedly surprised me. Very few people were explicitly rude to me; no one ignored my questions or requests. But few cared to make the encounter more comfortable and less frightening for me. No one asked me how I feel about my illness. I quickly learned that when doctors ask "How are you?" they mean "How is your body?"; that when an X-ray of my lungs is up on the screen and several doctors stand around it discussing my "case", they will not include me in their discussion even if I am standing right there. That they will not want to know how my life has changed because of my illness, how they could make it easier for me. This is one of the reasons I believe phenomenology's emphasis on the first-person experience of illness would be beneficial to the relationship between patients and health professionals.

Over time, things became easier in this respect. I became a familiar face at the respiratory department; the physiologists and nurses stop and say hello, ask me if I need anything. I learned my way around the hospital and became intimately familiar with blood tests, medical shorthand terms and the CT scanner. I learned to do my own research, find out what I need and ask for it. I found an experimental treatment that I decided to try. My consultant supported me, although it was a long shot. I asked to be referred for a transplant assessment. I was duly referred and had all the information I needed. My respiratory nurse, Ros Badman, was always available and friendly and on more than one occasion spared me days of anxiety and uncertainty. My consultants seemed to be doing what they could to look after me and things became less impersonal.

But I still remember the two appointments I had the day after my diagnosis. My family, a sad five-person convoy, made its way in and out of buildings, up and down lifts, in and out of offices. We shook hands with grey men sitting behind cluttered desks. They both said the same: there is no treatment for LAM; only observation. "And then?" I foolishly asked. The older of the two shrugged his shoulders. "You will have to have a lung transplant," he said. I felt like an idiot for asking such an obvious question. I didn't need to ask what would happen otherwise.

I observed my parents, both of them medical doctors. Suddenly they were sitting on the other side of the desk. They were the ones seeking answers; begging at the doctor's door, saying "but surely ..." and "there must be *something*". Watching them turn from confident figures of authority to emotionally defeated parents seeing their sick child become sicker, made it clear to me that there is some unspoken severance, some invisible but clear dividing line, separating "us" from "them". I remembered the humility with which patients knocked on our door, asking to see The Doctor. Suddenly we were one of those families.

Perhaps doctors and nurses need the dividing line to sustain their sanity in the harsh world of illness, pain and death. Perhaps no one can witness sorrow and offer empathy on a daily basis. Perhaps the medical world as we know it would not be sustainable under such a shift. I began to ask about this dividing line and its function. I interrogated my friends who were medical doctors. I began to speak to medical audiences. There seemed to be no consistent line among those I spoke to; it was an aspect of medical work that was entirely up to the individual. How health professionals dealt with their patients' decline and death was not discussed in medical training, they said. The emotional labour involved in seeing people at their toughest, most intimate hour was not acknowledged.

But there did seem to be a consensus: we will speak to one another in a dry, emotion-free way, patients quick to mimic the doctor's sanitized way of speaking about their illness, about their body. A further decline in lung function is "disease progression ". Pain is "a symptom". Fear cannot be spoken about. I wonder why this is considered the right way to speak about illness. I also wonder if I should have been offered help: counselling, contact details of the British Lung Foundation. I wonder what the first few months of my illness would have looked like if someone had asked: how are you coping? Can we help you in any way? What have you lost through your illness? Can we help make up for it somehow?

My revolt against the attitude towards illness that is common in the medical world is not a sentimental one. I am not suggesting that health professionals' precious time be wasted on feel-good chatting. But I do often wonder if the encounter must be so impersonal, so guarded. Could some genuine care be introduced to the exchange? I often wondered why it is that not a single medical practitioner has ever said they are sorry to hear I am ill. Such a banal social convention: I am sorry to hear about your illness. Why does this convention cease to apply as soon as one enters a hospital ward or a doctor's clinic? I wonder, for example, if a dedicated physiologist could do my breathing tests, rather than a new person each time. Wouldn't such a system actually be more efficient, if the physiologist knew my name and details, if we had a routine way of doing things? And if I felt comfortable with her, surely that would save rather than waste time?

In my many hospital appointments, scans and referrals I have seen both friendly and unfriendly health professionals. I have been talked down to, as if I were deaf or mentally deficient. I have spoken to doctors who did not wish to hear my opinion or take the printed information (helpful in the case of a rare disease such as LAM) that I provided. And when I began to give talks on illness, my position was never quite clear to the participants. Is she a patient or an academic? It somehow seemed to make a difference.

These experiences led me to believe that with the best will in the world, something is lacking in medical training or in the application of this training in practice. Surely a few hours could be spent surveying the profession, the concepts underlying it and the ways in which illness impacts patients' lives. How easy it would be to spend a few hours discussing not medicine, but the philosophy of medicine. Not "how to" or "what to do", but "what do you think about" health and illness? Are the concepts sufficiently defined? Is naturalism a satisfactory approach? What is the experience of illness like? What are the important aspects of the patient–physician interaction? Why do patients still complain of feeling alienated, objectified?

A few hours spent on such reflection could alter some of the conventions guiding medical practitioners today. A new discipline, medical humanities, and the philosophy of medicine address these questions and issues. Medical humanities courses examine the history and cultural aspects of medicine, how medicine is experienced and described outside the medical world, for example in literature and poetry. Philosophy of medicine scrutinizes the concepts and ideas underlying medical practice. Many medical schools around the world offer medical humanities courses as options for their students but it takes a long time for new ideas to trickle down to the consulting room.

The core idea of phenomenology is pertinent here. If healthcare practitioners devoted more time to understanding the experience of illness, much of the misunderstanding, miscommunication and sense of alienation that patients report might be alleviated. Phenomenologically inspired medicine would become a genuinely human science, where each term illuminates the other. One way of developing such understanding is by enabling the medical practitioner to have first-hand experience of the patient's world.

A recent attempt to provide medical practitioners with such an experience uses virtual reality simulators to afford the experience of angina. By strapping a thick belt around the chest and creating pressure there, the pain and constriction characterizing this condition are simulated. Other virtual reality simulators provide the experience of various kinds of visual impairment. Other machines simulate the perceptual and motor experiences of a person affected by a stroke, dementia or schizophrenia.

Another phenomenological avenue uses Merleau-Ponty's holistic approach to personhood in nurse training in order to shift the focus away from disease and symptom alleviation to a more holistic view of the experience of illness. Such work shows the great diversity in different patients' experience of illness, which demonstrates the need for a phenomenological approach to understanding the experience of illness (see Chapter 3). Other studies show that a phenomenological research methodology is successful in extracting and utilizing the first-person experience of the ill person. These are all reasons for medical practitioners to look to phenomenology for a different way of approaching illness.

So far I have spoken of first- and third-person perspectives. I juxtaposed the first-person lived experience of illness with the third-person observation of the ill person's body. I now want to suggest another perspective, that of the second person. Perhaps the I–Thou relationship – one person facing another – unmediated, ethically demanding, acknowledging each other's humanity, is also missing from the medical encounter. The twentieth-century Jewish philosopher Martin Buber (1878–1965) saw this relationship as the fundamental connection between any two people.

On Buber's view, human existence is a series of encounters and is dialogic in nature. This encounter can either be an I–Thou (one person genuinely encountering another) or I–It (a person encountering an objectified world) encounter. Communication and dialogue are two fundamental concepts describing the interpersonal nature of human existence. The I–Thou encounter brings out the universality, mutuality and authenticity of human exchange. It is a primary way of encountering another human being and the recognition of the immediacy and potency of the encounter. This encounter can of course be refined, structured and filtered through conceptual and cultural frameworks (turning it into an I–It encounter) but it remains a possible form of real dialogue.

I would like to suggest that some elements of the I–Thou encounter could be incorporated into the patient–health professional interaction. The immediate, non-objectifying I–Thou encounter is one in which "the parallel lines of relations meet", as Buber says in *I and Thou* (1971: 26). This is a possible relationship that would not require revolutionizing the

National Health Service. Nor would it require extensive training. It would take what some health professionals do naturally, and formalize it into a fundamental principle guiding the exchange between patients and health professionals. Speaking to patients as equals, showing fundamental human empathy and compassion, understanding that but for the grace of God it could have been you, are some components of such a relationship.

A phenomenological approach to illness has tangible benefits. It could improve the patient–health professional relationship by being an antidote to the objectification and alienation many patients complain of. In the past seven years I have had daily contact with other patients from the UK and across the world. The complaint that seems to appear near-universal in this context is: why am I not treated as a person? This complaint points to a certain culture within the medical world, of treating disease as a purely biological dysfunction. If disease is seen as a malfunction of a body part it (and the ill person) will be treated very differently than if it is seen as a world-transforming event, modifying the life-world of the ill person. A phenomenological approach would introduce the missing first-person perspective on illness and would enable health professionals to understand the transformation of the world of the ill person caused by the illness.

A phenomenological approach would clarify to the health professional what the impact of illness is on the ill person's life and it would address the asymmetry of the encounter. Addressing this aspect of the patient–health professional relationship may be beneficial to other issues, such as trust and compliance. Perhaps if the patient felt she was viewed as an individual by the clinician, she would trust her more and listen to her suggestions more attentively. This approach could also enhance interviewing techniques and ways of listening that could, in turn, lessen the danger of misdiagnosis. And finally, the patient experience, the way ill people experience their time in hospital or at the health centre, may be radically changed if they feel that their loss and the ways in which their world has become limited have been acknowledged.

LEARNING TO BE RUDE

My husband and I are invited for dinner at our friends' house. They are a lively couple and gourmet cooks and I look forward to the meal. A second couple arrives shortly after us, people we've not met before. We chat for a while and eventually sit down to dinner. Our well-meaning hostess tells the other couple that we just got married. "Oh, wonderful!" says the man. "Are you going to have children?" My heart sinks. The question drops straight into my core and sits there, heavy. My mind goes blank. "No," I say, in a tone I hope sounds stern enough to ward off any further questions. But he presses on. "Why not?" he asks. I am beginning to panic, but am not sure how rude I can really be to this man, a friend of our friends. I want to scream at him "because I am dying of lung cancer, you idiot". I think how illegitimate this question has become for our generation, especially for couples in their mid-thirties, who are either having trouble conceiving or decided not to have children, often after a painful process. I also think how personal, how damn personal, the question he is asking is. I want to hit him. I want to vanish from the table. A storm rushes through my mind. I say, "Because we can't".

I hope that my lack of enthusiasm for the subject will kill the conversation or at least make it awkward enough for someone to bring up a new topic. But the man is relentless. He presses on: "Is it something physical?" By now it has become obvious that he is in an unusual frame of mind, something between sadism and sociopathy. I don't feel I should or can answer that question, one too many, but by no means the last. My husband, equally embarrassed, mutters something about adoption. The man picks up on the topic and conversation about adoption ensues. I shrink further back in my seat.

I am sitting at the corner and would have to squeeze past two people to get away, undermining my intention to leave the table discreetly. After about ten minutes the man says another thing, a horrible thing that makes me want to cry. He tells about his neighbour who has children of her own but also fosters children. "I am worried," he confides in us, "that they will be a bad influence on my children."

This is it. I can no longer cope. My mind goes blank and my existence shrinks into a bitter ball of resentment. I think about him, his three children, his obliviousness to anything, and I mean anything, outside his world: his inability – a striking feature of many healthy people – to conceive of the lives of others. I think about how much I wanted and still want children and how little I allow myself to think about it. I think about the lovely, amazing, beautiful life we could have had. I think about a birthday party for a four-year-old I saw a few days earlier in the park. The children were sitting in a circle, their excited little faces glowing, singing Happy Birthday to Max. I think about how little I know about the world of parents, my friends who have children, and who always, even when complaining, seem so bloody content. I think about decades of joy and involvement with the wondrous world of children, all taken away from me, unfairly, bluntly, with no compensation, no explanation and little to take its place.

I am no longer listening to the conversation. I am thinking dark, horrible thoughts but at the same time trying very hard, and barely managing, not to cry. My ears are buzzing in a way that hurts but also blocks out outside sounds and voices. But the man – yes, this same Horrible Man – suddenly stops talking and says to my husband, "I don't think your wife is very happy about this conversation". All eyes turn to me. I know I've lost the battle against my own tears. My husband turns to me and it is his tenderness, his eagerness to make things OK, that clinches it.

I cry. I've had my fill. I leave the table and go to another room and sit quietly in the dark. I sob, but my sobbing is insincere. It is more of a gesture, a cursory marker of my sorrow. I know that I can't really cry for everything that has happened to me. The tragedy unfolding itself at my expense cannot be cried for. Like every other time I have cried since I became ill, I feel that this sobbing is not even the tip of the iceberg. I also know that the only way to deal with this iceberg is to leave it alone. To carry on with my life and the projects that are still possible while muffling the existence of the iceberg.

I also feel like an idiot. I know I should either return to the table or go home. I can't sit here forever. My husband comes in and asks me to come back. "Please," he says, "come and eat your food. It's OK." But I can't. I am still overwhelmed by the experience of being humiliated, violated, pierced by the questions of the Horrible Man. I am also overwhelmed by my fragility, my vulnerability. Why didn't I say something aggressive and unpleasant to him, to make him understand that I will not tolerate his questions? Why didn't I have something witty, but also sharp, up my sleeve? Why am I the one sitting here in the dark, feeling terrible, while the Horrible Man is chatting at the table?

I continue to sit in the dark. Eventually I will go upstairs to the bathroom and wash my face. I will stare at my sad face in the mirror, observe the red, swollen eyes, a small stain on my shirt. I will flush the toilet and switch off the light and come back downstairs. My hostess meets me in the corridor. She is in tears, apologising, saying that the Horrible Man has mental problems; he suffers from depression. She didn't know he was ill. I begin to see the comic side of it, what a wonderful scene this would make in a Buñuel film. The terminally ill childless woman meets an unhinged Horrible Man at a dinner party. I can't help but laugh.

I have since relived the exchange at the table and discussed it with friends and family until it stopped being interesting. I don't know whether the Horrible Man and his wife have thought about what happened that night. But something else has emerged from this: the need to be on your toes, to be ready for an attack – deliberately ruthless or naively insensitive – that will inevitably ensue: an attack on you as a person whose life trajectory is different; as someone who looks different; as a stigmatized individual whose condition is feared and denied by those surrounding you.

I cannot be responsible for the emotions and fears my oxygen and illness bring out in others. I cannot help them face their fears brought to the fore (albeit temporarily) by my illness. And people do respond in an alarming variety of ways. What I learned from this encounter is that it is only a matter of time before another person – a drunk on the street or a rude teenager – will say something that will bring tears to my eyes, something that will make the true horror of my condition appear in broad daylight, my carefully constructed but fragile defences crumbling before it.

A few months earlier I met another Horrible Man. He was a partner of an acquaintance of my husband. We were in a pub garden, having a drink before dinner. When sitting down, I appear entirely normal. I do not need to use oxygen and am not particularly breathless. So he didn't notice anything unusual and we weren't formally introduced. We got up to go for dinner at a nearby restaurant. We walked out, and while standing outside the pub, I put on my oxygen cannula, fitting it to my nostrils. I try not to do that indoors, where people may notice me. But outside, in the dark, I prepared to start walking, switching on the oxygen cylinder.

The man turned to me, standing perhaps ten feet away, lifted an accusing finger to my face and said, "Ugh, look. She's wearing nasal specs!" I said nothing, a result of shock and naivety on my part. His wife turned to him sharply and I could hear her whispering. The word "cancer" was the only one I could make out. The man appeared deflated. I did not speak to him; he did not apologise. I did not speak to his wife; she did not apologise either. No one else heard, or appeared to have heard, what he said. We had an oriental meal, sushi and fishcakes. I enjoyed myself and also enjoyed the fact that the Horrible Man said he was tired and left before we sat down. Was it remorse that drove him away? Probably not. Either way, I am sure that many years from now, these two men will bear no trace of the events that have been etched into my mind, making me paranoid, gruff and defensive in conversations with people I do not know well.

As a result of these kinds of encounters I learned to be rude. I learned to reply curtly to people who ask, out of vulgar curiosity, what it is that I am carrying. I learned I am not under any obligation to engage in conversations about my illness or my oxygen. Sometimes, when someone asks me something, I simply let out an ambiguous "Mmm". My husband learned the same lesson, and has several times chased people down the street, forcing them to apologise for thoughtless off-the-cuff remarks. We became wise, cynical, alert. We learned to be rude.

THE SOCIAL ARCHITECTURE OF ILLNESS

These two brief sketches provide an opening into understanding the transformed social world of the ill person. They describe encounters with strangers, encounters that are always, for me, in the shadow of the oxygen mask, in the shadow of my illness. But the changes in the ill person's relationships extend also to friends, family and spouses. In what follows, I would like to rethink some of these relationships in the light of illness.

The change in self-perception discussed in Chapter 1 is mirrored by changed social perception. How is the ill person seen by people in different encounters? How are ill people perceived by strangers, colleagues, acquaintances? Even more important is the role of friendship and the strains placed on it by illness. The experience of betrayal and disappointment, the threat illness poses to intimacy, and fear of the diseased body all impact on our relationships.

The transformation is most visible and damaging in the ways it hampers the ill person's social activities and interactions. It is seen in the ill person's difficulty in participating and reciprocating (e.g. if they tire easily, or are unable to cook), awkwardness around the subject of illness or disability, falling out of step with healthy people's activities and interests and personal factors such as anger and depression. These alone could cause severe damage to a person's social world. But there are others: the inability to ask for or give help, the loss of words and, harshest of all, friends who stay away because they do not know what to say.

Visible illness or disability often becomes the elephant in the room. It is seen as something that is not to be commented on or mentioned by polite people. But at the same time the condition challenges the normal interaction and makes *not* talking about it difficult, sometimes impossible. People often feel they ought to say something but are not sure what to say, or how, or when. They feel they should censor their expressions and self-reports, so as not to offend the ill person. The result is a general sense of discomfort, being ill at ease and unable to transcend the social barrier created by the illness.

When I was a graduate student I tutored an undergraduate student, a young woman who was blind. I was immensely struck not only by how difficult everyday life was for her, but also by the social isolation she experienced because of her blindness. When I took her shopping, salespeople spoke to me, as if she were deaf. When we went to the cinema, people stared at us, stares she could not see but I am sure could sense. When we went to the health centre, the receptionist began each sentence, directed at me, with, "Can you tell your colleague ...". When I took her to student society meetings, no one spoke to her. Her blindness was not only a physical affliction but a substantial social hurdle.

The visible signs of illness may arouse feelings of disgust, pity, anxiety or curiosity, which cannot be addressed in a routine exchange. It is difficult to find the right time and words to express these feelings. I witnessed many attempts by people to offer encouragement and support, to express admiration and caring. The striking feature of these attempts was how difficult they seemed for the well-intentioned healthy person. I can conjecture, from comments made by some of these people, that much more was left unsaid. At yoga class I often see people looking at me, and I imagine that they may have liked to say something appreciative, as some do, about coming to an advanced yoga class with oxygen. On the few occasions that people did speak to me, they said things like: "I admire your courage"; "I read something you've written in the paper"; "I was once very ill".

They often qualify their remarks by saying "I know this is a cliché", "You must have heard this a million times", "I know this is silly, but ...". I am sad that people feel they need to qualify such inherently helpful comments. This recognition of the ill person's coping is supremely important to her, to her feelings of acceptance, to her confidence. I always wish that people would say these things outright, because they acknowledge the myriad of terrible experiences – both physical and social – an ill person goes through.

There are additional problems facing an ill or disabled person in their social interactions. There are practical problems, such as being unable to participate in social events such as walking, dancing or drinking. Everyday activities have to be modified or sometimes given up if the condition does not enable the ill person to take part in them. The ill person can feel she is slowing the others down or hampering the natural flow of events, because of her special requirements and limitations, merely by being present. This, in turn, leads her to give up attending some events and a vicious circle begins.

There are also novel social issues that arise from the illness. For example, the ill person may feel shy about meeting new people because of the awkwardness created by the illness. She may feel the need to explain her condition and go into personal details. She could feel nervous about leaving the house and going to unknown territory, where the number of steps, wheelchair access, or the location of the nearest toilets is unknown. She possibly will not have the energy to participate in some activities, or fear that it would take too much effort.

A social architecture of illness mirrors the geography of illness discussed in Chapter 1. In the same way that distances increase, hills become impossible and simple tasks become titanic, the freedom to go out into the social world and improvise, to act and interact, is compromised. A new world is created, a world without spontaneity, a world of limitation and fear: a slow, encumbered world to which the ill person must adapt. All people fear this issue arising through ageing. In illness this new, bizarre, alienated world emerges, sometimes overnight. This is a world of negotiation, of helplessness, of avoidance. It is an encounter between a body limited by illness and an environment oblivious to such bodies.

The way the environment is arranged, the social environment included, makes it hard to negotiate while ill. Ill and disabled people invent a myriad of tricks, strategies and coping mechanisms to override the constraints inflicted on them by the environment and by the invisible background norms that govern the lives of all of us, ill or healthy. But the pressure is always there. I shall never be able to board a train, walk down the street or smile at a stranger in a way that would be unfettered by my illness.

The phenomenological idea of the transparency of health is illuminating here. According to Sartre, the healthy body is transparent: it simply does as you bid it and requires no special consideration or reflection. In illness, that transparency is lost. Similarly, the transparent and natural way in which we engage in social interactions suddenly becomes cumbersome, weighed down by unspoken doubts and discomfort, and the effort required for genuine communication becomes greater. The social impact of illness is the loss of this transparency and immediacy of social interaction.

This transparency of the body, of social ease, can be characterized more generally as a transparency of well-being. Well-being is the invisible context enabling us to pursue possibilities and engage in projects. It is the condition of possibility enabling us to pursue aims and goals, to act on our desires, to become who we want to be. But the spatial and temporal possibilities that characterize health are altered in illness, as we saw in Chapter 1. What I would like to emphasize here is not only the curtailment of spatial possibilities, but the abrupt descent of limits on to a world previously larger, freer, more open. These limits not only restrict physical movement, but inflect existential possibilities. It is not only physical possibility that suffers in the hands of illness. It is ways of being and ways of being-with that suffer.

THE ELEPHANT IN THE FRIENDSHIP

I remember well the time around my diagnosis. Friends rang, worried, confused. They asked questions about my diagnosis, my prognosis, the meaning of this long word: lymphangioleiomyomatosis. I had little to say back. I didn't want to know about my illness and I didn't want to repeat the grim, sparse information I did have. I sat at home, unable to talk or move, my hand constantly held by my parents and my sister. My sister and I went to the video store. We took out light-hearted films: Woody Allen's *Match Point* was one. There was a feeling of waiting for something, of anticipation. Every time a film ended we sat in silence, not knowing what to do. Every time the phone rang I cringed. No, I didn't want to speak to anyone. I had nothing to say. And, sadly, some of my friends had little to say too.

This silence marked the beginning of a new era in many of my friendships. Many people found it hard to communicate with me. Some said they "couldn't handle it"; some talked to others rather than to me, finding solace in the solidarity they felt. Some never got in touch. Only a handful of friendships were able to contain the blow. My friend Eran came for dinner the day after my diagnosis. He cooked, we laughed, ate a hearty meal comprising what was in the fridge: eggs, salad, vegetables. An air of abnormal normality characterized that informal meal, the first supper of a new era. Another old friend, Sharon, left her young son behind and came to spend the night. She stayed for two days but we didn't talk about my diagnosis. They reacted instinctively to my diagnosis: they came and sat with me, through the awkwardness and tears, sharing the burden of bad news.

With many of my friends silence remained the status quo. With many the fact of my illness is never mentioned and I have come to experience it as something highly secretive, grossly inappropriate for conversation. My dirty little secret. I often wonder how my friends perceive it. I don't know because we hardly ever talk about it. I presume some don't think about it often. Maybe others do not want to meddle with complex affairs. Perhaps those who haven't seen me in a while cannot imagine what has become of my body, of my life.

I distinctly remember two humiliating email incidents. One came from an old friend who was a graduate student with me. He heard I had returned from Australia, where I had been living for a couple of years, and got in touch. "How are you?" he wrote. "Send me your news. Let's get together soon." I wrote back telling him about my diagnosis, about having LAM. He never replied to this email. Another friend, in a similar context, also heard I had returned to the UK. He, too, emailed to get in touch. He did reply to my email saying "God, how terrible. I'll write more soon." He, too, has kept silent.

I can only guess what my friends and acquaintances think, from their distance, created by time, busyness, geography. But I can count on one hand the number of intimate conversations I have had about my illness with friends. Most of them did not and will not, out of propriety, fear, or loss of words, ask me about it. The status quo seems to be this: don't talk about your illness and we won't talk about our health, our healthy children, our pleasingly predictable lives. We won't talk about how everything worked out just fine for us, give or take a difficult labour, a premature baby or a divorce.

This bitterness in me has nowhere to go. It has no place, no name. It is *verboten*. The strict limitations on what I may or may not say even to my closest friends manoeuvre me into a more socially palatable position: being courageous. How brave I am. How uncomplaining. How cheerful in the face of a heavy, sometimes unbearable load. First I am set up in a social context that forbids me from talking about my illness. Then, when I turn to other topics, I discover the social reward: I am seen as brave, graceful, a good sport. Didn't she take it wonderfully, isn't she coping just marvellously? That is how you are seen once you conform to the demands and expectations of society:

once your "sick role" (as Talcott Parsons called it) is validated by those around you.

I paraphrase Erving Goffman, when I say: being a good ill person, a good patient, is conforming to the expectations of the healthy, not to be offended or polluted by your illness. When you begin to get praise for your behaviour, then you know you have achieved the status of a conformist. An ill conformist conforming to the demands of the healthy majority, who cannot, will not, wills not to see the fate that awaits us all.

And so my illness remains the elephant in the street, the cafe, the office. The elephant in the friendship. And perhaps we would have all been better off if my illness did not present itself as a taboo, something arousing pity and terror, the emotional components of tragedy according to Aristotle. I sometimes think that what is tragic about being ill is this silence.

This loss of immediacy and spontaneity affects both old and new relationships. New relationships are sometimes more difficult to form. The hesitation, the thought of being off -putting or appearing strange, the thought of being judged harshly as a freak – a young woman carrying oxygen, tubes sticking out of her nose – may make the ill person less forthcoming and less friendly than before. Old relationships, especially the less intimate ones, also suffer. Friends may not know how to keep up with the changes, how to ask for information that may be sad or sensitive, how to discuss the illness with the ill person. Friends who remain supportive need to overcome their own discomfort and fear of saying the wrong thing.

And sometimes they do say the wrong thing. I remember telling an old friend about my illness and how much I deteriorated in the past year. She recommended breathing exercises to me. "Try them," she said, "you might gain a bit of time". When

I told her I lost nearly 50 per cent of my lung capacity over a short period, she looked at me and said: "so if you lose another 50 per cent next year ...". It took me a long time to regain my composure that day. I never discussed my illness with her again.

These encounters made me think of the extreme discomfort elicited by illness. People feel terror, embarrassment and mortification when they are confronted with illness. Some respond to these emotions by confronting them, others by fleeing. Some overcome the initial sense of panic and surprise; others run a mile, from me, from illness, from themselves.

When I became aware of the possibility of lung transplantation, the life-saving procedure restoring health and giving life, I began to ask friends and acquaintances if they were on the organ donor register. It seemed so easy, so important, to register that I was certain everyone already had. To my great surprise, many people said they did not. "Oh," they said, "I just never got round to it." Having heard this reply over and over from people in their forties and fifties made me realize that it was not their busy schedule preventing them from registering. It was the squeamishness involved in thinking about their premature death.

Like those mortified by death, those who do not feel a responsibility to reflect seriously on organ donation also run away. But this evasion comes at the expense of many people who die needlessly because of this squeamishness, this refusal to engage with one's mortality and duty to others. I never heard anyone say that they would refuse to accept an organ if they needed it. It is the need to reflect on the dire circumstances that make organ donation possible that they shy away from. Having seen both the avoidable tragic suffering and death inflicted by the shortage of organs for transplantation, as well as the health and joy of those who were lucky enough to get an organ on time, I think that this evasion is not only foolish but also immoral.

This social awkwardness has a flip-side. It is also true, for me, that some people are damned if they do and damned if they don't. If they ask questions, I feel uncomfortable, as if they are prying. If they say nothing, I think they are selfish, self-centred, oblivious to my plight. If it is difficult to talk about illness, it is especially hard for ill people. But what I learned from my illness is that in times of hardship, grief and loss, there is no need for original, illuminating phrases. There is nothing to say other than the most banal stuff: "I am sorry"; "This is so sad". Saying this – and listening – are the best ways to communicate with ill people. Or so I believe.

Within the family, things are more complicated still. The transition from an independent person, an autonomous adult going about her business, to someone who needs help and special consideration can be devastating not just for the ill person but also for her family. Her parents may find it incredibly hard to see their child ill. Siblings may find the experience overwhelming, confusing. And children may secretly accuse themselves of causing the illness. I have heard many accounts of difficulties arising as a direct result of illness, including being deserted by one's spouse. In some cases the breakup occurred shortly after the diagnosis of the illness. A particularly striking story was of a woman who, upon her diagnosis with LAM, sat with her husband in the doctor's waiting room, waiting to discuss her prospects with the doctor. Her husband said he couldn't cope with it, and left her there and then.

Why do all these negative changes take place? Social exchanges normally occur between people who wish to be perceived in a particular way: to be admired and liked, to be seen as funny, original or attractive. The presence of an illness or disability is problematic in two ways. It radically curtails our ability to control what others think about us and it places the interaction, at least initially, in the shadow of the illness.

When someone is ill or disabled in a way that can be immediately perceived by others, they may feel highly exposed. It feels as though people can see through you; intimate details become the first thing a stranger sees about you. Instead of choosing what you disclose about yourself, you become a passive vessel of information provided through your own betraying body, a body that cannot keep a secret. A stranger takes a cursory glance at you and already knows so much about what is sensitive, intimate, painful. And yet, you know nothing about her. The asymmetry makes it even more difficult. This is the prelude to any interaction with a stranger. As a result, not surprisingly, the ill person may be reduced to her illness. Or she may anticipate this response and react defensively. For passersby and strangers I am "the woman with the oxygen", memorable for her deficiency.

There is a philosophical significance to this feeling of complete and defenceless exposure. One might think that in such a situation something that is perceived as an internal, intimate detail about you becomes external. But is this a correct description of the situation? Should the illness or disability be considered a secret, something I choose whether to disclose? This seems to be the wrong way of thinking about it. And this is because of a broader metaphysical error: the thought that there are two distinct realms, the internal and the external, that are completely distinct. In fact, if we return to embodiment, we can see that there is no distinction between the two realms. We are embodied; our consciousness or our self is one and the same as our body. Therefore there is no "inside" that is un-embodied.

Illness or disability is not an external mark of indiscretion. Nor is it a foreign and detachable part of the ill person. It is intimate and personal, yes, but at the same time it is a visible, objectively perceived feature of my body. This demonstrates clearly Merleau-Ponty's view of the body. His view goes beyond the alleged distinctions between the external and the internal, between objective and subjective points of view. The body is something that is both highly personal, highly subjective, and also exists as an object within a natural order of things. On this view, my body does not betray me by exposing my secret, but embodies my situation as an ill person.

So how can we negotiate the gap between the healthy and the ill? Do I contribute to the growing distance between me and some of my healthy friends? What do some of my friends do that makes it possible for us to talk about my illness? I think about positive examples to learn from. I think about my friend Catriona, a fellow philosopher, who has been persistent in her desire to know, to understand what I am going through. I think about one particular conversation we had. I was on a train heading to Edinburgh, just after having been to a meeting of LAM Action, the UK LAM charity. My heart was full of pain: the pain of seeing dozens of women with LAM, the pain of realizing that this is my life, this is how things are; the terror of thinking about becoming sicker, like some of the women there; the envy towards those who were only mildly affected by the disease. She knew I had been to the meeting. And so she phoned me, and stayed on the phone for a very long time. Green fields and sea views whirling past, I told her some of my fears and sadness.

I think about my neighbour Sarah, who walks my dog every morning when my breathing is at its worst. I think about her honest questions and her desire to help. I think about John, a physician-cum-philosopher from New Zealand I met at a conference. We had a magical, explosive exchange, full of bitter honesty and brutal medical facts. He had a cold so I spoke to him from a safe distance, but when I said goodbye my eyes filled with tears. He said, "It's been an honour to talk to you." And for once I believed it was true.

THREE

Illness as dis-ability and health within illness

There was nothing gradual about my diagnosis. In the morning I was still happily navigating the world as a healthy person although I got breathless easily. That afternoon I learned I had LAM. The next day I was told that my prospects were poor, that there was no treatment for LAM and that I would probably die within ten years. I had to get used to the idea that my life will be spent in increasing disability and breathlessness. That I will not be able to have children. That there is little point in making long-term plans. That I am unlikely to enjoy my pension. All the rules that governed my life until then have been radically broken and nothing, nothing, remained the same. I had to overhaul all my plans, expectations, goals, projects and horizons. Most importantly, I had to rethink my idea of a good life.

I was confronted with an objective fact: I was ill. My illness had no treatment. There was nothing for me to do but sit and watch. The possibility of being passive in the face of my own devastation was unreal to me at first. On hearing the news my first question was: so what do I *do*? I remember the radiologist's evasive answer: "I don't know. I only diagnose, I don't treat." Did he know there was no treatment? Was he trying to get out of a sticky situation? I don't know because I never saw him again and, despite being my father's colleague for many years, he never asked after me.

My first reaction was simple. OK, I thought, I'm ill. I should get treated. When I learned that there was nothing I could do to prevent my lung function from declining and that at the end of my illness lay death by respiratory failure, the extent of my tragedy became clear. How could it be? How could there be no treatment? I was young, healthy, I looked after myself: how could there be nothing for me to do? There is always *something*.

The "monitoring" and "observation" the doctors offered seemed to me like a morbid joke. Do they mean they will plot my decline on a chart and watch me die? I wasn't prepared for that. I was hit by a rare disease, an orphan disease: a disease no one knows about other than a handful of patients and their families and a few doctors who became involved with LAM. There are only about a thousand women diagnosed with LAM around the world, although the estimated number of women with the disease is much higher, as it is difficult to diagnose without a CT scan. Very few people have heard of it. There are hundreds of similarly rare diseases no one has heard of. Unless, that is, you have one.

A few days into my diagnosis I realized I couldn't do anything about it. The decline of my lung function was not within my control. The usual "virtuous = happy" equation did not apply in my case. I did all the right things. And yet here I was, gravely ill with LAM. Passivity was on the cards because there was nothing I, or anyone else, could do. Or so I thought.

The full name of the disease – sporadic LAM – made me feel a little better. It didn't happen because of something I did. It wasn't my fault. It was sporadic. Out of every 400,000 women, one will have LAM. I was the one. There wasn't any drug, any exercise, any operation, any diet that would make me better. People who recommended detox diets and acupuncture made me shake with rage. My lungs are being destroyed by a complex, barely understood cystic process and they are telling me to eat fruit.

I was even angrier with those who suggested that it was "all in the mind". On their account, there was something in the way I think, something about my psyche, but most importantly, something within my control, that made me ill. I was ill because I was "thinking negatively", because I "didn't want to get better". One person suggested that my negative thinking was affecting my body at a cellular level: that my morbid thoughts were helping LAM cells flourish in my body. Although mostly made in a friendly spirit, the actual significance of these suggestions is downright hostile and I took them as such.

The thought that was truly novel for me was this: I will never get better. All the usual rules that governed my life – that trying hard yields results, that looking after yourself pays off, that practice makes perfect – seemed inoperative here. It was the first instance, for me, of unconditional, uncontrollable failure. No matter what I did, I would only get worse. The inevitability of decline was the only principle governing my life.

Being good failed, and so I turned to denial. I just won't think about it. I will ignore my illness and carry on as before. This worked for a few months, but as my condition deteriorated and my illness became more invasive, more pervasive, I could not ignore it. I had to start using oxygen. I had to put up with the stares on the street, the nose bleeds, the breathlessness. I had to slow down. I had to ask for ground-floor rooms for my teaching. I watched with horror as my lung function halved within a year. I didn't just watch it, I experienced it. I looked at a hill I could cycle up easily two years ago, with difficulty one year ago, and turned around, knowing full well I couldn't cycle up it at all any more.

I still remembered doing everything – walks, trips, yoga, running, football, kick-boxing – without oxygen. The memory was alive not only in my mind, but in my body. My body rushed forwards, only to be halted by a force stronger than it, chained to its dying lungs. My movements were those of a healthy person, jerked back by my lungs' inability to supply the necessary oxygen for the most trivial everyday movements: hanging the washing, gardening, walking. I tried and tried and tried and tried, but I only got worse.

The first lesson I learned was how to give up. I realized that no matter how hard I try, I will not get fitter. That no matter how hard I blow into the tube, my readings are on a one-way trend, downwards. That even though I liked to pretend to myself I was just like the next person, I was not. And so I stopped trying. I surrendered, accepted, began to make room for my inability. I will never get better. This became my starting-point for everything else.

A second lesson was about internal states. Against the objective horror of my illness I cultivated an inner state of peacefulness and joy. I was surprised at this, as this ability to be ill and happy, to be gravely ill and yet feel so normal, was not something I expected. I don't know what caused this response but I thought about it like this: I have no control over this illness but I have full control over my emotions and inner state. I cannot choose to be healthy, but I can choose to enjoy the present, embrace the joyful aspects of my life and train myself to observe neutrally my sadness, envy, grief, fear and anger. The emotions washed over me. I observed them. By not fighting, by just letting them be, they became less mine, more tolerable.

When I was in my twenties I went to India with my good friend Sharon. On our first night there I woke up in the suffocating monsoon heat. Our ceiling fan had stopped working. I shook my friend saying: "Wake up! The worst of all has happened." This phrase, so comical in this context, remained a joke between us for many years. But now the worst of all has happened. And I was distinctly surprised by how it felt. I was surprised to learn that other things that had happened to me previously had hurt more. There was a limit to how much sadness I could experience and once this limit was reached, I bounced back, even if not entirely successfully.

So what is the worst that could happen? I asked myself. The worst that could happen is that I will die. So screw it; I will die. What's the big deal? Everyone dies; I am an instance of this general law. We wouldn't want to say that a good life is simply a long life. Surely what counts is how I lived my life, long or short. And so, I told myself, if I die, then that's that. After thinking about it and dreading it for months on end I got tired. I stopped thinking about my premature death because it no longer seemed interesting.

"What if?" and "if only ..." were two to avoid. The game of "what if a cure was found" and "if only it were the girl next door" is the most futile of all. It is me. There is no other reality. I live here, in this world, not the imagined "what if" one. I quickly learned not to play this game. Once I stopped playing it, I began to notice that many healthy people played it to their detriment. I observed people around me making up excuses for their unhappiness or just accepting it, complaining about trivial things, reliving the past for decades on end, refusing to take responsibility for their well-being, for their lives. I began to see how many sources of unhappiness there are; mine was only one of many sad variations on a theme. I learned to discern excuses from reasons, minor mishaps from crucial events and self-indulgence from true accommodation of inner limits. As the ancient philosopher Epicurus famously did on his deathbed, I learned to enjoy happy past memories rather than lament their passing.

I learned to live in the here and now. I learned to distinguish really bad things (a further decline in lung function) from unimportant ones (wrinkles). Small things stopped bothering me because when compared to having LAM they would deflate and disappear. Ironically, by really having something to worry about my mind was cleansed from many sources of anxiety. So many things didn't matter. Only one thing really mattered. And that one was beyond my control. I learned to respect two things: that the laws of cause and effect governing the universe may generate suffering over which we have no control and that everything, including myself, was ephemeral.

I had to change the way I thought of myself. I had to convince myself that my existence or non-existence is not so important. A movement away from vanity mirrored this gradual movement away from an ego-centred approach. Both were hard to achieve. I had to recognize that I was no longer able to do many things; that I have no control over what happens to my body. From thinking of myself as a young, healthy, promising life and body I had to start thinking of myself as fragile, damaged, unable.

BEING UNABLE TO BE

Heidegger characterizes human existence as "being able to be" (*Seinkönnen*). Human existence is characterized by its openness, potential, ability to become this or that thing. This underpins a powerful picture of human life: one can become what one wants (apart from certain physical and temporal limitations). If I wanted to be a polar explorer, I would have to train, build up my strength, learn to navigate and so on. Eventually, I would join a polar expedition and fulfil my plan, achieve my goal.

The plans and aims we have connect our present actions (for instance studying navigation) to a future view of ourselves as being able to do a particular thing (navigate to the North Pole). Present actions have meaning in virtue of being part of a project that is forward-looking, futural. I do something now in order to become something in the future. This definition of the human being is best understood by Heidegger's notion of projection. Projection means throwing oneself into a project, through which a human being's character and identity are enacted. If my project is being a teacher, I project myself accordingly by training to be a teacher, applying for teaching positions and so on. This, Heidegger claims, is the essence of human existence: the ability to *be* this or another thing, to assume a role as a teacher, a polar explorer and so on.

This view of the human being as becoming, as able to achieve her aims, as constantly changing according to the project she pursues, is appealing in many ways. It credits us with the freedom – and the responsibility – to become what we want, to shape ourselves and our lives in a way we find fulfilling: to transcend our present self with a future self that is more developed, more able. This progressive view of the person sees it as constantly changing, growing, developing. As Merleau-Ponty says, echoing Husserl's and Heidegger's approach, being in the world is not a matter of an "I think" but an "I can" (1962: 137).

But what about the other part of life, the one in which we become gradually *un*able to do things, unable to be? What about decline and insufficiency? In the physical sense, this aspect of life is undoubtedly there. Someone may be an athlete for many years, but eventually her body declines and she is no longer able to be an athlete. Does Heidegger's definition exclude this important aspect of life, that of decline, inability, failure to be?

When ill or ageing, we become unable to do some things, perform particular roles and engage in certain activities. This poses a problem for Heidegger's definition because it shows it excludes important human situations. In some illnesses, especially mental and chronic illness, a person's ability to be, to exist, is radically changed and sometimes altogether curtailed. Certain projects must be discarded and sometimes, as in severe psychosis, the possibility of having a project at all becomes impossible. Could Heidegger's account allow radically differing abilities to count as forms of human existence? How flexible are human beings? How much can we adjust our projects and plans in the face of ill health?

I suggest that Heidegger's definition is often understood too literally and his characterization of existence as "being able to be" needs to be modified in two ways. First, the notion of "being able to be" must be broadened to include radically differing abilities. Secondly, "inability to be" needs to be recognized as a way of being. Heidegger's definition can be given an interesting twist if we think about being unable to be as a form of existence that is worthwhile, challenging and, most importantly, unavoidable.

We should interpret the notion of "being able to be" as broadly as possible. It should include cases in which the smooth operation of the body, its assistance in carrying out plans and projects, is no longer there. Current projects may have to be abandoned and new projects created. These new projects have to be thought of in light of new limitations and therefore arise within a restricted horizon. But radically differing abilities all count as abilities to be. Take a person in a wheelchair, someone with terminal-stage cancer, learning disabilities, or Down's syndrome – all of these are ways of being that differ in some respects from the mainstream. But they should nonetheless count as human ways of being. Perhaps the outcome of applying Heidegger's notion of "being able to be" to cases of illness and disability is an acknowledgment of the diverse ways in which it is possible to be and the ways in which human beings differ from one another.

The opposite of being able to be is of course *not* being able to be; but this presupposes that the notions form a dichotomous system. We can replace this dichotomy with a spectrum of abilities to be. There are other modes of being able to be that are excluded by this dichotomy. Being partially able to be, learning to be able to be and rehabilitating an ability to be are a few examples. The ability to be that characterizes human existence is territory to be experientially explored and developed, rather than delimited through this opposition.

We can easily find positive examples of this. Stephen Hawking may have wanted to be a footballer or pianist, but because of his illness he was unable to pursue these projects. Instead he had another project, being a physicist, and has become extremely successful at it. It is true that many projects that might have seemed attractive to him were closed off because of his illness. But even within a contracted horizon of possibilities, there is still some choice.

We can also think of processes such as rehabilitation from drug use or stroke; learning to be able to enjoy life after severe depression; being only partially able to walk, hear, see or talk and so on. None of these conform to Heidegger's definition in the full sense, but if we understand ability to be more flexibly, we can account for such cases.

Furthermore, in cases of ageing, disease or disability we need to acknowledge an inability to be as a way of being. One way of thinking about ageing and illness is as processes of coming to terms with being unable to be. As coming to think of one's existence as more reliant and less independent, more interlinked and less autonomous. The inability (or the altered ability) to be and do is the flipside of Heidegger's account of being able to do. For some individuals it is there throughout life, as in cases of chronic illness or disability. For all of us it is there as a late stage in life, the stage of ageing and decline. Inability and limitation are part and parcel of human life, just as ability and freedom are. By introducing the notion of being unable to be as an integral part of human life we can move from seeing ability as positive and desirable to seeing it as part of a broader, more varied flux of life.

Being unable to be is not an independent or context-free concept. It has to be seen in relation to being able to be. An inability to be is a modification of an ability to be that is lost. Being unable to fly, or unable to breathe under water and so on are not examples of being unable to be. Otherwise the concept would have nearly endless examples and we would be more unable than able to be if all were taken into account. It is a *lost* ability or an ability that is never achieved viewed against a background of a common capability. Being unable to be is therefore intimately linked to an ability to be, and vice versa. Being able to be is not infinite, unlimited. It is a way of existence that is granted temporarily, for a number of years, and is never guaranteed, never certain. It is a fragile, transient gift. The notion of inability to be reveals this aspect of being able to be.

Even in cases of extreme physical disability there is always a freedom of thought, imagination, emotion and intellect. Freedom and imagination can enable even those who are unable to be in one way to be in a new way. One contemporary example of overcoming a physical inability to be is that of the virtual. New personae and images are created in a virtual space, where physical inability is irrelevant. Other spheres of imagination are helpful too: the worlds of literature, film and art may not be physical, corporeal, but within them the imagination can roam free and liberate the unable body, albeit temporarily, from its inability to be.

Acknowledging an inability and learning to see it as part of life's terrain are important lessons that illness can teach at any age. This knowledge enables the ill person to embrace the unable self as part and parcel of human existence. By having a more balanced view of life as combining ability and inability, illness can be greeted with increased acceptance and become less disruptive. In addition, the ability to accept an unable self can be the first step towards answering a pressing question for ill people: given that I am ill, can I still have a good life?

CAN I BE ILL AND HAPPY?

Can I be ill and happy? This question quickly became central to my life as an ill person. Since my illness has no treatment and no immediate prospect of a cure, I had to learn to live with it. Moreover, I wanted to live well with it. Was it possible for me, or for any other ill person, to be happy and live a good life?

For many centuries philosophers have been debating what a good life is. One way of answering this question is to list the conditions for well-being, for example, freedom, health, access to social goods, self-fulfilment and so on. Although health is, of course, taken to be an essential element of the good life and a necessary condition for happiness, philosophers have, for the most part, overlooked a particular issue. This is the question: what happens to the good life or to happiness when health is permanently absent? This is a fundamental question because ultimately the vast majority of people die of some kind of illness and a significant minority spend their lives chronically ill or disabled.

The question has gone virtually unnoticed within philosophy because of the taken-for-granted nature of the body, a privileging of the mind as the locus of intellectual pleasure and a denigration of the body in some Western traditions. Accounts of the good life recognize the importance of health and of quality healthcare to well-being. But they fail to consider the possibility of permanent absence of health, as in chronic illness. It is not enough to say that health is an essential ingredient of the good life, when it is often absent and current medical knowledge is unable to restore health. What we need to ask is whether it is possible to have a good life despite ill health.

As we tackle this question, we shall continue to use a phenomenological approach. This approach provides us with a relational account of the ill person, which includes her daily activities, goals and interaction with an environment and a social world. Once the change in these dimensions is understood *vis-à-vis* the illness, we shall be able to begin to answer our question. Viewing illness from this perspective will also enable us to seek a remedy for the negative changes. But in order to expose these changes we need a view that enables a complete description of the ways in which the life and world of the ill person changes. As I hope has by now become clear, such a view is offered by phenomenology.

The introduction of a phenomenological approach to illness is not intended to displace but *augment* the naturalistic approach. Phenomenology distinguishes the lived from the biological body, but does not reject the latter. Of course the biological body is central to any conception of illness. What I am contesting is the possibility of understanding illness – and the question of well-being in illness – only through the biological body, while ignoring lived experience.

In previous chapters we talked about how illness splits apart the biological and the lived body. Instead of the flawless correspondence between our objective body and our lived experience of it, in illness the biological body behaves oddly; it exhibits strange symptoms and becomes unpredictable. The transparent function of the biological body is gone and it is now the subject of anxious attention and medical scrutiny. The otherness of this body is brought to the fore; it may behave erratically or become unrecognizable. Objective facts about the biological body cease to tally with lived experience. To return to the example of anorexia, one could be underweight (objectively weigh 6 stone) but have the lived experienced of an overweight person ("I feel fat; I need to lose weight"). In this case the lived and biological body have fallen out of step, become alienated from one another.

Illness distances us from the biological body, which becomes alienated and erratic, the source of pain and disability. The lived experience of this body becomes painful, unstable, treacherous. This distance from the biological body is not normally available to us while we are healthy. Illness (as well as other kinds of physical alteration) removes the body's transparency and problematizes it. The body becomes the focus of concern, a source of pain and fear, and thus becomes a problematized body in two senses. It is the source of practical problems and concerns and it is also metaphysically unstable because its previous position and relationship with lived experience have broken down.

As we shall see towards the end of this chapter, the movement between the biological and lived body also has a creative potential. Ultimately a new relationship between the two can emerge. This new relationship reflects not only the practical complexities of illness but also the philosophical complexities of the body-subject: the body that is both a material object and the seat of subjectivity. Phenomenology gives us the tools to think about this relationship between the lived and biological body, between the body as subject and the body as object.

From this perspective we can see how illness is not simply a problem in an isolated physiological body part, but a problem with the whole embodied person and her relationship to her environment. Because the lived body is not just the biological body but one's contextual being in the world, a disruption of bodily capacities has a significance that far exceeds that of simple biological dysfunction. This change is essential to an account of the good life and is substantially disrupted in illness. It is not just a body function that is disrupted. Rather, one's entire way of being in the world is altered.

For example, one component of the good life that is undoubtedly disrupted in illness is the ability to assert oneself or to pursue one's goals freely. Phenomenology captures the important relationship between this ability and the body. A change in the body and in physical and perceptual possibility transforms subjectivity itself. The possibility of agency is inherently linked to the ability to perform actions and carry out activities that promote one's goals. The loss of physical or mental ability carries with it a reduction in the ill person's ability to pursue her goals because even the most abstract goal requires bodily action. Whereas it is normally taken for granted that the body is a healthy functioning enabler contributing silently to the execution of projects, in illness the body comes to the fore and its pain and incapacity directly affect the agency of the person.

The phenomenological concept of illness contains the relation of the ill person to her world, or what I earlier called the geography and the social architecture of illness. This includes spatial and social relations, which were discussed in Chapters 1 and 2, and temporal relations, which will be discussed in Chapters 4 and 5. The geography and spatial relations of someone with poor lung function are entirely different from those of the able-bodied person. Subjective notions of distance, difficulty and so on are modified as bodily abilities change. While an able-bodied person perceives stairs as a means for getting somewhere, for a person with lung disease the same stairs present an insurmountable obstacle. This is not just a local disturbance of the ill person's activity, but a fundamental alteration of the way she engages with her world.

Similar changes can be seen with respect to inclines or visibility, for example, depending on the particular condition affecting a person's interaction with her environment. The world of the ill person is changed, and notions of distance, time and level of difficulty go through a fundamental transformation. In order to give a complete account of illness we have to consider this transformation. We also need to provide an account of the transformation for practical reasons, in order to seek remedies for the changes and losses involved in illness. And most importantly for the subject of wellbeing, the transformation has to be fully described and must be taken into account when asking whether happiness is possible for ill people.

The social world, so crucial to happiness, is also transformed. Many disabled and ill people report difficulties maintaining their social life because they are no longer able to participate in shared activities such as work, trips and sports. The reciprocity characterizing most social relations is lost if the person is housebound. Also, ill people sometimes voluntarily withdraw from social situations that may compromise or embarrass them. Relationships come under new kinds of pressure when the autonomy and independence of the ill person are compromised. Old friendships must change to accommodate the illness or eventually wane. New friendships are now formed in the shadow of illness.

The ability to interact freely within a social context is lost. When you enter a room with oxygen, everyone will register this fact. What do they do? Do they talk to you normally? Do they ask why you are wearing it? Do they avoid talking to you because they are not sure what to say? Some level of awkwardness is unavoidable. The illness becomes a social issue as much as a personal one. This is apparent in the work of many mental health and AIDS charities, whose goal is to reduce the stigmatizing of these diseases. Physical conditions are stigmatized in a different way but create social difficulties nonetheless. These issues, too, are crucial building blocks of well-being and happiness.

Before we can turn to the question of well-being in illness, we must first highlight problems that are specific to long-term or chronic illness. This is important, because if an illness is temporary, it normally does not impact one's long-term happiness. But chronic or long-term illness poses the challenge of how to seek satisfaction and well-being despite, or with, the illness. In the case of chronic illness, the naturalistic approach further exacerbates the situation. On the naturalistic view, chronic illness is perceived as a physiological problem and therefore only physical suffering or loss of physical function are taken into account.

But there are other, grave, losses. The loss of agency, when the ill person is no longer as independent or as effective as she was, is one. The loss of productive function, if she has to leave work, or work less, or ask for help with some things, is another. The loss of ability to participate socially if she can no longer socialize in the same ways as previously is yet another. And finally, there is the possible loss of financial status because of healthcare expenses or being unable to work. These losses lead to great changes in the ill person's life, but are not recognized as part of the illness and are not addressed by the healthcare team.

Moreover, because chronic illness is sometimes treated within a framework of care designed for acute illness (such as a broken leg) that is sudden and temporary, the result is fragmented care. Care may be based on an acute illness model, and lack continuity. This may lead to incomplete information because of lack of continuity in the care and to overburdened caregivers, if the system is not geared towards chronic illness. Finally, many chronically ill people report feelings of isolation. The patients may not know other patients with the same illness, have no access to support groups or may not know where to find further information on their condition. Encountering a different nurse and doctor with each visit to the clinic exacerbates these problems. All these are, of course, factors affecting well-being.

We usually think of illness as a temporary abnormal condition that should be rectified so as to return the ill person to normal life. But in chronic illness this aim cannot be achieved. Chronic illness is often misconceived within the framework of acute care, with its view of illness as temporary disruption of self, rather than as a condition causing continuous losses. What is called for in the case of chronic illness is attention to the transformed world of the ill person, with more thought given to how to ameliorate the losses caused by illness.

Long-term illness or disability redefines the relationship of the person to her world, and moreover transforms this world by altering and limiting it. As embodied persons we experience illness primarily as a disruption of lived body rather than as a dysfunction of biological body. But medicine has traditionally focused on returning the biological body to normal functioning, and has therefore worked from within a problem-focused, deficit perspective that ignores the lived body. Within this approach, the experience of the ill person is measured in negative objective parameters, that is, how ill or impaired she is, while the lived experience of illness, which varies tremendously from one person to another, is overlooked.

A phenomenological approach can provide a framework for incorporating the experience of illness into an account of the good life by providing a rich description of the altered relationship of the ill person to her world. This altered relationship is a natural part of the life cycle and as such must be part of any account of the good life. Such a description takes illness to be a relational concept, one that must include the ill person and her physical and social world. What happens to the well-being of people who are seriously ill? Are they less capable of having a good life?

HEALTH WITHIN ILLNESS

So far we have treated health and illness as two opposed, mutually exclusive concepts. I spoke of "the healthy" and "the ill" as if they are two distinct groups. Susan Sontag describes this dichotomy at the beginning of her essay *Illness as Metaphor* with these words: "Illness is the night-side of life, a more onerous citizenship. Everyone who is born holds dual citizenship, in the kingdom of the well and in the kingdom of the sick" (1978: 3).

I suggest that health and illness are not dichotomous or separate kingdoms. Moreover, in the same way that episodes of illness can occur within health, an experience of health within illness is a possible, if often overlooked, phenomenon. Health within illness is a concept developed in the 1990s, to express the diversity of illness experiences. Using a phenomenological approach and its emphasis on first-person experience, researchers interviewed patients with chronic illness, to assess the impact of illness on their lives and how they respond to it. Their findings were nothing short of surprising.

Different studies show large variation in the meaning and impact of illness and in the coping mechanisms developed in response to it. Some claim that the experience of illness can promote personal growth through awareness and transformational change. Others note that for some chronically ill people illness became a tool of self-discovery and a fundamental source of later self-development. Many reports suggest that chronically ill people experience health subjectively and individually.

These perspectives acknowledge the possibility of health within illness, regardless of the person's physiological condition. They also seem to mirror accurately chronically ill or disabled people's reports. A notable phenomenon found in many studies is the lack of correlation between objective health (the biological body) and subjective well-being (lived experience). Studies show that subjective well-being is by far the domain least affected by chronic medical conditions. There is substantial evidence of individual variation in well-being that is not accounted for by age or disease condition.

Surprisingly, many chronically ill people rate their health as good. In a study by Stuifbergen, 73 per cent of interviewees, all of whom were living with a disabling condition, rated their health as good or excellent. Moreover, the individual definitions of health varied considerably among interviewees. While some defined health as "never to be sick or taking medication", others used "being able to take care of myself", or "enjoy life each day", revealing the multidimensionality of the concept of health. Similarly, in a survey conducted in Canada, 60 per cent of respondents suffering from health problems rated their health between good and excellent.

These findings require a shift in the way we think about health and illness. We need to move from seeing them as mutually exclusive opposites, towards a continuum or a blend of the two, allowing for health within illness in people who are objectively ill. A second shift is required, away from an objective deficit-centred health assessment towards giving more weight to subjective first-person reports or to lived experience. Both shifts are achieved by moving from a purely naturalistic view of health and illness to one that incorporates phenomenological insights.

This shift is also expressed in the change "from cure to care", moving away from a model of disease and cure, to a model of care that promotes health and healing for people with chronic illness. Ultimately the aim is to change healthcare practices by reconceptualizing the health–illness distinction and offering a broader perspective on health and health experience.

Researchers influenced by phenomenology interviewed chronically ill and disabled people and noted themes provided by the interviewees, such as honouring the self, creating opportunities, celebrating life and transcending the self. These themes were developed from participants' own words and therefore accurately reflect their illness or disability experience. As one quadriplegic interviewee said, "I can live life to the fullest, even if I have no physical ability, I can still live life to the fullest because where I am living, life is from within" (Lindsey 1996).

Other studies present different findings. Many studies report feelings of loss, chronic sorrow, frustration, guilt, anger, loss of connectedness and struggling with change. Some found negative attitudes, a diminished sense of self-identity and a grieving process . The difference lies in the focus of the studies. Studies that focused on the experience of health within illness, rather than on the experience of illness, led to radically different results. There seems to be a hidden dimension of health within illness that only emerges when researchers ask patients explicitly whether they experience episodes of well-being or of feeling healthy within their illness.

This brings to light another limitation of the medical approach. Because the focus is on disease and a negative deficit approach is applied, the positive experience of health within illness remains unacknowledged within medical practice. For example, many medical test results are presented as a range indicating the patient's position in it. Lung function is expressed as a percentage of predicted capacity, but is not matched to a subjective sense of functionality or well-being, that is, what the individual can *do* with this lung function. The actual function – how active the individual is, what sort of physical activity she does and so on – is marginalized in medical records, despite a well-documented discrepancy between objective parameters and actual functionality.

Studies have repeatedly found that chronically ill people's experience of quality of life is subjective and individual. Surprisingly, happiness does not correlate with objective health (with the exception of pain and incontinence) and many substantial chronic illnesses seem not to affect the ill person's subjective sense of wellbeing. For example, having had cancer does not seem to affect one's subjective well-being.

The philosophical value I see emerging from these studies is the plurality of responses and attitudes, which show great variation between individual cases and also within each case. If we think about how illness appears in someone's life, this should make sense. An ill person can have good and bad days, acute episodes and periods of stability. Moreover, no condition is experienced identically by two people. Also, there is a genuine ambiguity in the meaning of chronic illness and the concept is too broad to be stable. If illness is part and parcel of life, and on a continuum with health, then our experience of it will be as diverse as our experiences of health or of life in general. In other words, it would be difficult to generalize this experience.

There are many different kinds of chronic illnesses and many different grades of severity within any illness. Chronic illness could strike someone when young, in which case it is a constant accompaniment to her life, or it can appear at a later stage in life, when illness is less surprising. Chronic illness may cause depression or it may be incorporated into a productive, happy life without impacting on it too heavily. These differences must be taken into account if we want to understand the experience of illness. But despite the variation, an important point remains valid. When studies ask about illness, the interviewees' responses focus on illness. But if they ask about health within illness, new and more positive dimensions of illness emerge.

ADAPTABILITY AND CREATIVE RESPONSES TO ILLNESS

So what are the positive mechanisms enabling health within illness? The two I focus on here are the ability to adapt to new, more limited capacities and the creativity that emerges in response to novel challenges. I use the term "adaptability" (distinguishing it from the evolutionary biologists' term "adaptation") to refer to the behavioural flexibility enabling ill or disabled people to adjust their behaviour in response to their condition.

In the case of illness, two notable features emerge. First, the adjustment is to a change not within the environment but within one's body. The change is experienced immediately and unpredictably, leading to an experience of alienation from one's body, which has become uncanny, unfamiliar. This makes the change unmediated and more intimate than any environmental change. Secondly, the introduction of instruments to overcome bodily limitations could be thought of as a hybrid of body and tool.

What I mean by "adaptability" is not just psychological adjustment, coming to terms with illness or acceptance. Adaptability is much more diverse. Adaptability can appear in physical, psychological, social and temporal domains. The changes occur simultaneously in several domains and often blend into each other. The most notable feature of these types of adaptability is that they arise as a response to changes in one's own body rather than to a novel environment. This endows adaptability with a highly personal dialectic nature, one of disruption followed by rapid response. The tension between the body as active and passive, subject and object, capable and unable, presented with an obstacle and overcoming it, is present in adaptability. It is not a smooth process but a series of dialectic encounters of a body with an environment, of a demand with failure, and of failure with the need for modification.

Another feature is that the change is negative. The change reflects deterioration in some bodily capacity. I do not wish to focus on the negativity of the change but on the highly personal and creative dimension of the response to it. This focus goes hand in hand with the notion of health within illness.

In face of a diminished bodily capacity, ill people must find or invent solutions to novel problems and challenges. The ill person may need to adapt her walking speed, gait, time allocation, level of physical activity and so on, depending on the limitation created by the illness. I adapted to my breathlessness by finding circuitous routes to the shop to avoid walking uphill; I pause several times while climbing up a flight of stairs and allocate more time for everyday tasks such as gardening or having a shower. I have also become more aware of the physical aspect of minor tasks and my body's new responses to exertion.

Some types of physical adaptability can be viewed as automatic or subconscious. For example, the physiological response to breathlessness and oxygen desaturation is panting. This response is not controllable and may lead to a sense of loss of control, an experience of alienation from one's own body. The transparency and taken-for-granted nature of the body are now replaced with an acute sensitivity to the body's demands and the limitations it places on the ill person. In illness you must become more attuned to your body, more attentive to its signals, demands and limitations.

Although they are responses to a negative bodily change, these kinds of adaptability also have a creative element. Finding a new way of performing an old task, given an altered set of capacities, is challenging; successful performance leads to a sense of achievement. For example, a physiotherapist told me that patients who completed a course of physiotherapy and regained some of their lost capacities report a high level of satisfaction and improvement in quality of life. Being able to improvise and create new ways of compensating for a lost ability shows the plasticity of behaviour and the human capacity to adjust to change.

Psychological adaptability is of a different nature. Some sociologists of health such as Simon Williams have described illness as biographical disruption. The disruption is of taken for granted assumptions and behaviours (especially focusing on the body, which no longer "passes us by in silence" as Jean-Paul Sartre says) and of the explanatory framework (raising questions such as "why me?"). The response to the disruption comes as mobilization of medical, financial and cultural resources.

The view of illness as biographical disruption is illuminating. Becoming ill creates a need to find meaning for a new narrative: the narrative of health that has now been disrupted by illness. The ill person seeks an explanation for her suffering and limitations; she must renegotiate old habits and relationships; she may need to give up an identity and she must create a new approach to both her present and her future. There is a lot of adaptive work in making these adjustments and the ill person sometimes has to re-tell her life story ("everything was going well until ..."). The old narrative with which she described her life may no longer be suitable and a new narrative must be weaved and endowed with meaning.

One question that arises is how life can continue to have meaning, to make sense, when one has been struck by illness. Issues of fairness and desert often give rise to bitterness, envy and despair. A deprecating, damaged self-image may arise and will be reflected in the ill person's narrative ("bad things always happen to me"). A new or modified narrative will have to address these issues, and to tie together the old narrative with the new, disrupted one.

The mechanisms of coping, normalization, strategic mobilization of resources, accommodation and denial all play an important part in psychological adaptability. Changes to a person's sense of self and identity are common and both patients and researchers use concepts such as enduring, struggling and disruption to describe the experience of illness. An interesting exception is a study of East-End Londoners' attitude to illness, entitled *Hard Earned Lives*, which notes their cheerful stoicism and pragmatism towards illness. The explanation of their unique upbeat approach is provided by the author, Jocelyn Cornwell (1985). These people, she says, view illness as a normal part of life, and use terms such as "normal illness" and "health problems which are not illness". This attitude makes it easier to cope, as those affected by illness do not feel unfairly plagued by ill health. It is possible to imagine similar acceptance on the part of older patients who feel they have lived long and happy lives that are now winding down.

Even within the psychological strain created by illness some positive adaptive responses are notable. For example, an adaptive theme appearing prominently in some studies is that of gaining control of an altered life direction. Other adaptive themes include confronting loss, struggling for normalcy, reformulating the self, transcending suffering and courage in the face of adversity. These adaptive themes demonstrate the continued exploration of self and creation of meaning against an adverse background.

Adaptability is also expressed on a social level. Here examples of patient advocacy groups, patient activism and patient-driven research demonstrate a positive adaptive response. Other types of social adaptation may be finding new activities the ill person can share with others to replace group activities that are no longer feasible in light of the illness. For example, a woman with a chronic respiratory condition who could no longer go on long hill walks has replaced this group activity with Tai-Chi, a more gentle form of physical activity that can be done in a group.

Temporal change, which will be discussed in the next two chapters, is notable. More time is given to each activity, which in turn may cause that person to begin experiencing herself as older than her years, as "useless" or as more disabled than she is. As her movements slow down, the difference of pace between herself and others become evident. She needs to translate healthy time ("it's a ten minute walk"; "it's very easy to get there") into her own, idiosyncratic tempo.

The unpredictability of the course of illness leads some to adopt a perspective of living in the present and refraining from looking towards the future, making long-term plans or having rigid goals. Illness may restrict the ability to imagine future scenarios, causing a further shrinking of the ill person's world.

Within these restrictions, ill people can adapt by developing an ability to live in the present. They can think less frequently about the future and miss their healthy past less intensely. Expectations of a particular future can be modified, or replaced by other, more modest expectations. People often talk about a capacity to understand the fragility and transience of life and nonetheless appreciate life's goodness and value as a notable feature of illness. This rare insight is another creative dimension that is available through the experience of illness. As Sigmund Freud writes in "On Transience", "a flower that blossoms only for a single night does not seem to us on that account less lovely" (1985: 288).

What may seem to the nurse like a very short time in the waiting room is an agonisingly long period to the person waiting for test results. Waiting times, periods separating one consultation from the next, waiting for a phone call from a doctor, have often been described as the worst aspect of being a service-user of a public healthcare system. The patience required to be a patient demands of those who have less time (both because they may not live long and because being ill is time-consuming) to spend it in waiting: waiting for test results; waiting to see the doctor; waiting for the next appointment; waiting for the painkiller to kick in. Time spent in pain or discomfort is experienced as slow, tedious and enduring.

The medical team, on the other hand, may experience time as flying by while they are busy with work. As a result, the medical team and the ill person experience the temporality of illness differently. These different experiences of time contribute to the fundamental disparity between the objectively perceived, naturalistic model of disease and the lived experience of illness. This is another example of how introducing a phenomenological approach to medical training could reduce the disparity and assist the physician and the ill person to co-define the temporal dimension of illness and to understand each other better. A further aspect of adaptability and creative responses to illness and disability is that of the extended body. As Merleau-Ponty points out, external props such as a walking stick can become an integral part of one's lived body. When a tool such as a car or a musical instrument becomes intimately familiar and incorporated into everyday practices, it is no longer an external object but becomes part of the lived body schema. If my bike has a puncture, I experience the breakdown as impacting on my mobility, not just on the bike.

If we apply this idea to medical equipment we can see the use of a wheelchair, for example, as novel in two senses. First, one body part is used to compensate for the loss of function in another (e.g. using hands instead of legs to move). Secondly, an external artefact is used to compensate for loss of function. On Merleau-Ponty's account, the wheelchair becomes incorporated into the lived body schema and is no longer experienced as an external addition. He writes: "the blind man's stick has ceased to be an object for him, and is no longer perceived for itself; its point has become an area of sensitivity, extending the scope and active radius of touch, and providing a parallel to sight" (1962: 143).

This is an additional perspective on the body-as-tool view I presented earlier. Heidegger provides us with a sophisticated tool analysis, and argues that we only notice tools when they cease functioning, go missing, or get in the way. He calls these situations, respectively, the conspicuousness, obtrusiveness and obstinacy of the tools. We can think of our body as similarly conspicuous and obstinate when it no longer does what we expect it to. But it is also interesting to think about the ways in which a body's obtrusive malfunction can be ameliorated by adding a tool to it to create a hybrid. A blind man with a stick, a paraplegic in a wheelchair and a LAM patient with an oxygen cylinder are examples of such hybrids. Tools can help us overcome the conspicuousness of the ill or disabled body.

Earlier in this chapter I explained how illness creates a gap between the biological and lived body. But a second, reconciliatory stage is possible, when the biological body that was transformed by illness is reunited with the lived body. The reunification takes place through acceptance, new tolerance and even joy found in the altered, diseased biological body. The two are reconciled by appropriating the transformed body and integrating it into the experience of illness. This synthesis may take many years to achieve, but is another creative achievement possible in illness.

Happiness and a good life are possible even within the constraints of illness. But their uncovering requires a new set of conceptual tools (such as health within illness, adaptability) and a metaphysical framework that gives precedence to the experience of illness and to the embodied nature of human existence (phenomenology).

What are the consequences of this approach? What actual changes can it bring to healthcare practices? I believe that medical practitioners would do well to adopt a broader, less exclusively naturalistic approach if they want to assist their chronically ill patients maintain their habits, activities and goals. The naturalistic approach provides a limited, biological picture of illness and therefore fails to help us understand the experience of illness.

As physicians themselves report, the phenomenological approach is threatening to many practitioners' outlook. Their outlook requires strict separation between their and their patients' lives, an objective, sanitized language and lack of engagement with social and emotional aspects of illness. The naturalistic approach provides protection from the personal, while the phenomenological approach requires such engagement. If physicians want to help their chronically ill and disabled patients, the questions they should ask are: how has illness or disability changed your life? What aspects of it affect you the most? How can those effects be made up for? These kinds of phenomenologically informed questions open a space for the creative adaptability that can enable a good life even with illness. They give due place to the context, experience and relations of the ill person and assist in maintaining a modified but nonetheless rich texture of life even without medical resolution of the disease.

FOUR

Fearing death

One of the many questions facing ill people, or at least those with poor prognosis, is: how should we face death? Of course, this question is one that every person faces, in some way, with the realization that their existence is finite. The knowledge of our mortality is a heavy burden, and many philosophers have claimed that cultivating an appropriate attitude towards it is the key to leading a good life. Some, such as Socrates, Cicero and Montaigne, thought that practising philosophy is learning how to die, because the practice of pure thinking separates the soul from the body. As Montaigne writes in his essay "To Philosophize is to Learn How to Die": "study and contemplation draw our souls, somewhat outside ourselves, keeping them occupied away from the body, a state which both resembles death and which forms a kind of apprenticeship for it" (1993: 17).

Montaigne provides a second, more compelling reason for the view that philosophy prepares us for death: "all the wisdom and argument in the world eventually come down to one conclusion; which is to teach us not to be afraid of dying". Philosophy's practical or therapeutic role is to prepare us to meet death with equanimity. Preparing for death is thus also a lesson in how to live well. As Socrates says, those who pursue philosophy rightly "study to die" and "to them of all men death is least formidable" (*Phaedo* 67D). Philosophy can help us live well by helping us to cultivate an appropriate attitude towards our finitude.

Death is a central problem for humanity, especially for ill people who face concrete and imminent concerns regarding it. But how can phenomenology tell us anything about death? If phenomenology is a study of lived experience and death is the end of life and of experience, how can there be a phenomenology of death? The answer is that we are not dealing with death, but with our *fear* of death. And so phenomenology will serve us well when we try to understand and analyse this fear, when we reflect on the kind of experience thinking about death is. Phenomenology is not interested in analysing actual death, but our *relationship* towards it. And as we shall see in this chapter, there are important reasons to believe that death will not harm us, but our fear of death certainly will.

Heidegger argued that knowledge of mortality is a constant presence in life, and renamed human existence "being towards death". What is human existence, he said, if not a limited stretch from birth to death? Human existence is marked by finitude and limitation, and those who ignore this fact are engaged in a futile pursuit, trying to escape the inescapable. To understand life fully, Heidegger argues, you *must* understand yourself as finite.

The ancient Greek philosopher Epicurus (*c.*341–270 BCE) took a different stance when he argued that the fear of death is irrational. "So death, the most frightening of bad things, is nothing to us; since when we exist, death is not yet present, and when death is present, then we do not exist. Therefore, it is relevant neither to the living nor to the dead, since it does not affect the former, and the latter do not exist" (1994: 29). So long as you are alive, death is nothing to you. And once you are dead, you are no longer there to feel anything, fear included. Life and death are mutually exclusive. If we think about it carefully, said Epicurus, we will realize that what we are really afraid of is not death but dying, the pain of illness and decay. There is nothing to fear in death itself, because death is a state of non-existence. It is incoherent to say that you fear not existing, because non-existence is simply not being: how can it be feared? Epicurus thought that those who fear death were confused and should use rational arguments, like the ones he provided, to overcome their fear.

Which of these two views is more compelling? In some ways, Epicurus is right. If I do not believe in the persistence of my mind after the death of my body, what have I to fear? I am not afraid of non-existence. But on the other hand, there are other things to regret in death. Like the grief of others, missed opportunities and having to take a different course in life than the desired one. Let us consider both views in more detail.

BEING TOWARDS DEATH

Heidegger thinks that death cannot be regarded as irrelevant to life, as Epicurus seems to say. On the contrary, the only way for us to understand our existence fully is by seeing it as finite. Life, for Heidegger, is a constant movement towards death.

His formulation of human existence as "being towards death" captures the temporal essence of human existence. This temporal essence condemns us to be constantly propelled through time towards our own annihilation. This annihilation is not culmination or achievement. Unlike a fruit culminating in its ripe state, or writing a novel, which culminates in the final published book, human life ends in nothing. Human life lacks the final goal, or *telos*, that the ripe fruit and the finished novel have. We propel ourselves through life, but to what end? None, says Heidegger. There is nothing at the end. This is the fact we need to make sense of.

Our mortality is at once a key fact about human existence, a structuring element of human life and consciousness, and something that has no meaning. It simply is that way. We are finite. So what is the point of thinking about it? Why doesn't Heidegger embrace the Epicurean "if you can't beat it, forget it"? Because Heidegger thinks that although we cannot defeat mortality we are nonetheless obliged to understand ourselves as finite, to grasp our mortality, in order for us to be able to lead a good, or what he calls authentic, life.

Grasping our mortality is essential to being authentic because otherwise we may treat time and events within it inappropriately. If we do not grasp our finitude we may squander time, we may unthinkingly see ourselves as immortal and therefore respond inappropriately to situations. Most importantly we will not be able to conceive of ourselves as temporal, as existing in time, and so will misunderstand our structure. Self-understanding, for Heidegger, relies heavily on understanding ourselves as finite, limited creatures that are first and foremost temporal. We exist in and as time: we change with time, our past existence makes us who we are in the present and we make future plans towards which we project ourselves.

In order for us to understand the meaning of our existence we must first see our lives as a stretch from birth to death, or as essentially temporal. We must understand our non-existence on both sides, but especially our death, because that is where we are inexorably headed. When we have created an appropriate view of our temporal structure we can then fill it with meaning and authentic understanding can emerge. This authentic understanding relies on understanding ourselves as a unity of past, present and future, with each temporal mode affecting the others. This temporal unity of the three modes describes the structure of human existence (the contents will of course be different for every individual but these differences are contingent while the structure is essential). As Heidegger writes, "finitude is not some property that is merely attached to us, but is *our fundamental way of being*" (1995: 5).

Like Merleau-Ponty, Heidegger was a phenomenologist. He, too, thought that lived experience is the key to self-understanding. But a key feature of death when viewed from a phenomenological perspective is that we never experience our own death. Death is not a phenomenon contained within our experiential horizon nor is it an experience that we undergo. It is important to see that Heidegger is not offering an analysis of what he calls demise, the event that ends life. Heidegger is not interested in the event that transforms a living body into a corporeal corpse. Rather, his analysis focuses on the ways in which our existence is shaped by mortality and how life is, paradoxically, a process of dying. Heidegger does not provide a phenomenology of death but an analysis of being towards death, a phenomenology of mortal, finite existence.

Equally Heidegger is not presenting us with a phenomenology of the death of others. Although someone else's death could be a definitive and profound event for me, I am barred from experiencing their death. At most, I experience my loss. As Heidegger writes: "Death does indeed reveal itself as a loss, but a loss such as is experienced by those who remain. In suffering this loss, however, we have no way of access to the loss-of-Being as such which the dying man 'suffers'" (1962: 282).

So Heidegger's analysis provides us with a phenomenological account not of death, but of our relatedness to death, or in other words of our mortality. Because of death's inaccessibility it can only be related to by scrutinizing its effect on life. It is clear that Heidegger is not focusing on the moment of demise, which is experientially inaccessible (I may experience being ill or even dying, but not death itself). Rather, he focuses on the *anticipation* of death and on our existence as being towards death.

Heidegger therefore looks to life and everyday existence and examines the ways in which they are shaped and affected by death. So although death is external to experience and not an event within it, death influences everyday existence and our structure, which is now characterized as being towards death. In *History of the Concept of Time*, Heidegger writes: "the certainty that 'I myself am in that I will die' is *the basic certainty* ... insofar as I am, I am moribundus. *The moribundus first gives the* sum *its sense*" (1992: 317).

Death grounds our existence because it constitutes us as temporally finite. As Heidegger writes, it is "only in dying can I to some extent say absolutely 'I am'" (*ibid*.: 318). This gives a revised meaning to our temporal structure, first and foremost to the notion of projection, which we encountered earlier. As we said, we constantly project ourselves towards our future, by making plans, carrying out projects and choosing to pursue certain possibilities over others. In this projection of ourselves towards our future we also project ourselves towards death: the impossibility to be anything or to have any more possibilities. Death is the possibility of no longer being able to be. So our movement towards the future is a movement towards annihilation, towards *not* being able to be.

We press into the future by projecting ourselves towards our chosen possibilities. But this movement is not entirely free; it is bound by our past choices and actions, as well as by some of our given features: we are born into a certain family, place and culture. We are historically and socially situated in ways that are given to us rather than chosen by us. This is what Heidegger calls our "thrownness". So, for example, I cannot choose to become a Victorian actress. Nor can I choose to become a medieval nun. I cannot choose to spend my twenties any differently than I already have and the choices I made during those years will limit the possibilities I have open to me now. Some options are not available to me and the ones that are, are delineated by previous choices I made.

Ultimately human existence is described by Heidegger as "thrown projection". I am thrown into a world, a historical and political period, race, state, gender, family and background over which I have no control. In that sense my past is something into which I was thrown. But I am also constantly projecting myself into my future, choosing which possibilities to pursue and which to let go of. My future is something into which I actively project myself through my choices. Understanding human existence as thrown projection expresses the fact that the freedom to press into a certain possibility, the freedom to shape one's future, is not unbounded. It is limited by a contingent past and a finite future. The formulation of human existence as thrown projection expresses the idea of *bound* or *finite* freedom. The existential and philosophical task is, therefore, according to Heidegger, "to conceive freedom in its finitude", or to understand our bounded ability to choose within limitations.

This idea of bound freedom is further expressed in the fact that death is not an ordinary aim or event that we can project ourselves towards. While all other possibilities give us something to be, death is the closing down of our temporal structure. It is also different from other possibilities because it is unavoidable. While I can avoid certain possibilities and choose to steer clear of certain projects, I cannot choose not to die. "[E]veryone owes nature a death and must expect to pay the debt", as Freud said (1985: 77). This necessity generates a difference between my death and the death of others. All events, including the death of others, take place within my world and are therefore subsumed under my experiential horizon. But my death is the closure of my experiential horizon: it is the possibility of the impossibility of existence.

We not only have a finite structure, but are also endowed with the ability to conceive our finitude. One unique feature of human beings is our ability to understand that we are going to die and therefore to live as finite. Death is not only an objective fact structuring us as temporally finite. This fact is also reflected in the way we live our life. Death is therefore not only an external endpoint but bears internally on how we live and what kinds of projects and choices are open to us. Although we may not explicitly express it in this way, we always make plans within the horizon of temporal finitude. We normally make plans for – at best – the next few decades, not for the next millennia; we do not take on personal projects that cannot be completed within a human lifespan.

So death is a limitation that is expressed, among other things, in the types of projects we choose to pursue and plans we make. It is therefore implicitly present in our self-awareness and self conception. We can state this more broadly and claim that death is a constant accompaniment or condition of all events within life; it is always there as a possibility, regardless of what I am doing. Death is a constant shadow accompanying every action, even if we never think about it. Because it is constantly (if implicitly) present in life, the phenomenological project of understanding death is a demand to *understand life as finite* (*not* to understand the event that ends one's life). For us, death is an issue and a unique projection that we must somehow face. It is this relation to death that Heidegger explores, and the reason he focuses on being *towards* death, rather than on death.

But the projection towards death is a unique and problematic projection. Phenomenologically it is a projection towards something that cannot be experienced, as we just noted. It is a projection towards something that is *not*, towards annihilation. So being towards death is a projection towards obliteration. This makes our existence a continuous movement towards extinction, where death is a paradoxical culmination of existence. As Heidegger formulates it, death is *the possibility of the impossibility of any existence at all*: my world disappears with me, or at best becomes someone else's contingent past.

Whereas other possibilities give us an identity, something to be, death gives us nothing to be, no role or personal identity. If I choose to become a polar explorer this choice gives me an identity, a project: it gives me something to *be*. Death, on the other hand, does not give us anything to be, and in this sense it is not an ordinary kind of possibility. Rather, death destroys the human being, who is nonetheless compelled to constantly move towards it. Although the movement is certain, the time and manner of our death remain indefinite, so it is a constant accompaniment to every moment of our existence. We know death will be our end, but we do not know when or how. This makes it all the more important to our self-understanding.

Heidegger points out three features of death: it is *ownmost, non-relational* and *not to be outstripped*. The term "ownmost" indicates death's essential belonging to each individual. This characteristic singles out death as something that cannot be taken away from a particular individual or passed on to someone else. We each owe nature a death. In this sense death is different from other attributes or responsibilities that can be performed by or given to a different person. Someone else can teach my class for me if I am ill, or volunteer to donate blood instead of me. But death is my ownmost because even if someone else sacrifices their life for me, I still have to die myself.

The second characteristic, non-relationality, expresses death's individuating effect. Death singles each person out and severs her relations to others. Death severs us from friends and family, from relationships with other people, animals and things. Death severs our links to the world. It removes us from the network of relationships and meanings that make up our world. So when we face death we are confronted with our individuated, pure existence, separate from our world, friendships, and links to objects and people.

The third characteristic, not to be outstripped, is a combination of two further attributes, death's *certainty* and *indefiniteness*. Because it is certain, the threat of death hangs over us constantly. Because it is indefinite, that is, we do not know when we will die, we are constantly anxious about its arrival. As a result, we cannot overtake death, so to speak. We are not able to hold it fixed as we move through time, the way we can with other events. Take, for example, a friend's visit. We can wait for a visit planned for next week (it is ahead of us, in the future). When the week passes the day of the visit arrives. The visit then takes place (it is now, in the present). Once it has ended we can view it as a past event, an event that has already taken place and has been temporally surpassed. Death cannot be similarly surpassed. As long as I exist, it is always in front of me and always indefinite in its arrival.

Nabokov's 1935 novel *Invitation to a Beheading* is a literary exploration of this interplay between the certainty and indefiniteness of death. The protagonist, Cincinnatus, is a convict on death row, awaiting his execution. Cincinnatus is repeatedly tormented by false announcements of his impending execution made by his sadistic jailor. Cincinnatus says, "The compensation for a death sentence is knowledge of the exact hour when one is to die. A great luxury, but one that is well earned. However, I am being left in that ignorance which is tolerable only to those living at liberty" (1959: 14). Death is a possibility that is distinctively and only ever impending. It is something that each individual has to take on herself in every case, but that cannot be controlled, surpassed or temporally determined like other events.

We constantly move towards our death. We always *anticipate* death: death is always and only ever something that is yet to come. We never expect its actualization, because death gives us nothing to actualize. To anticipate death is simply to live as finite and as a consequence to understand our structure more fully. For Heidegger understanding our finitude reveals our ability to be, the ways in which we can exist as finite. And most importantly, within these different ways of existing, it uncovers the possibility of authentic existence.

Our relation to death is not something we should understand merely theoretically. Being towards death is an active practical position. Our way of being towards death uncovers the possibilities towards which we project ourselves, our particular movement towards the future. Ultimately, says Heidegger, when we anticipate death it frees us, because death illuminates all other possibilities as being part of a finite structure. Viewing ourselves as such a finite structure enables us to view our existence as a limited whole. This understanding, again, is not theoretical but practical, enacted. We not only *understand* ourselves as a finite whole but *exist* as one. When we understand ourselves as a finite whole we will have a better understanding of human existence, of what is possible for us and of what existence means.

There are two ways for us to respond to our mortality: authentically and inauthentically. We can choose to respond authentically to death, to live life with an appreciation of its finiteness. This attitude opens for us the possibility of engaging authentically with our existence, since we have now grasped it more fully *as finite* and have enhanced our understanding of ourselves as thrown projection or finite temporality. We can also flee from death and cover it up by ignoring death. Heidegger calls this attitude "inauthentic".

These two attitudes to death, authentic and inauthentic, are not merely philosophical or abstract. These attitudes underlie our everyday practical concerns and types of engagement with the world, because all our actions are performed within a temporally finite horizon. As a result no one is exempt from having some sort of attitude towards death, even if it is one of avoidance. Whether we assume an authentic attitude towards death by resolutely facing our finitude, or whether we flee from our mortality, we are always bound by death.

So what is the authentic life Heidegger proposes? Living authentically involves what Heidegger calls perspicuity, having a clear overview of one's life. And this in turn demands two things: understanding individual situations fully, and having a coherent grasp of one's temporal existence – past (birth), present and future (death). In order to gain perspicuity and become able to be authentic, it is essential that we grasp our death.

But in what way should we grasp our death? Heidegger seems to say that our awareness of death should be a constant accompaniment to any living moment. That any living moment should be understood as unique, irreversible and as bringing us closer to death. Because time only moves forwards and because we are temporally finite, each moment is an essential ingredient in our life's arc from birth to death. We must take responsibility for how we live because there are no second chances. I can never repeat today. I have chosen to live this day in a particular way and there is no turning back. Appreciating our finitude enables us to live each moment with full appreciation of its distinctiveness and the weight of responsibility implied by this view.

EPICURUS' FOUR-PART CURE

On Epicurus' view, the main obstacle to happiness is anxiety, which has many sources. But if we understand the irrationality of our anxieties and live by four simple beliefs, suffering and anxiety will disappear from our lives. He thought that human beings are ultimately rational and that the fears and anxieties that cloud our minds can be dissipated if we invest time and effort in thinking clearly about them. His view has been summarized by Philodemus, a later Epicurean philosopher, as the four-part cure (*tetrapharmakon*).

The four-part cure treats fundamental areas in our lives that cause us fear and suffering. It consists of the following:

> Don't fear God,
> Don't worry about death;
> What is good is easy to get, and
> What is terrible is easy to endure.

Fearing God (or the gods, in Epicurus' time) is one type of mistaken emotion we may experience. The gods, according to Epicurus, are not interested in punishing or rewarding us, nor do they run the world. The gods are rather like abstract, perfectly tranquil human beings, unconcerned with our affairs. At most they can be role models for us, because of their tranquillity and equanimity. Since they are not there to punish or reward us, there is no need to fear them. (I am glossing over certain subtleties; my aim is not to provide a complete account of Epicurus' philosophy, but to discuss the aspects of his philosophy that are relevant to death).

We have already seen why we should not worry about death, according to Epicurus. If death is a state of non-existence, there is nothing it is *like* to be dead. Being dead is not painful, boring, sad – or anything else for that matter. Being dead is the same as not being conceived. We are never sad about missing the times before we existed. We never think that pre-natal non-existence should trouble us. Similarly, says Epicurus, we should not be bothered by posthumous non-existence.

What we may well be bothered by is the thought of dying in pain or suffering prior to our death. Well, for that we have the fourth proposition: what is terrible is easy to endure. Epicurus famously suffered quietly through the terrible pain of kidney stones. He did not complain and thought that physical pain – which is terrible – is easy to endure with the solace of good friends and a glass of wine. In fact, he considered the day of his death the happiest day of his life, because his physical suffering was far outweighed by mentally reliving pleasurable past experiences. For us the situation is simpler. With the invention of powerful painkillers and sedatives, dying is no longer as painful as it used to be, although it can be degrading and frightening and pain is not always successfully controlled.

Finally, why does Epicurus say that what is good is easy to get? Surely many of us would like to be rich, famous or admired, none of which is easy to get. But Epicurus does not think that wealth, fame and admiration are good things. Truly good things are only those that come from within. These are things that cannot be taken away, unlike the external effects of wealth and fame. So being dependent on wealth, expensive wine or the admiration of strangers in order to be happy would make us vulnerable to the whims of fortune. If I depend on my wealth for my happiness and my wealth disappears with time, I am condemned to misery.

So what are the truly good things, the things that cannot be taken away? One of the main goals of Epicurean philosophy is to make us realize how little we need, how much pleasure we can get out of simple things and to assist us to cultivate the (pleasurable) belief that we will continue to possess these things. Epicurus also wants us to understand that cultivating unnecessary desires is ultimately self-harming, as the more we rely on external pleasures, the less self-sufficient we are. By chaining ourselves to pleasurable external things we ultimately risk pain (mental or physical) when these things are out of our reach. Epicurus is not an ascetic and does not think that luxuries are intrinsically bad. It is only their potential to cause us suffering in their absence that makes him suspicious of unnecessary pleasures.

Friendship and tranquillity are two things that are truly good and that do not (usually) depend on external circumstances. And tranquillity, in part, comes from understanding that there is nothing to fear in death, especially when we inhabit the point of view that Goethe called "the eternity of the present": being here and now, cherishing the present and not being preoccupied by plans and projects.

The four-part cure rests on certain metaphysical assumptions about the nature of human existence and human psychology. In order to understand why, according to Epicurus, we should not fear death, we need to look more closely at these assumptions and explain his moral and psychological view of human life, namely, hedonism.

EPICUREAN HEDONISM

According to Epicurus, all feelings divide into pleasure and pain. Everything else is reducible to how it feels to us: whether we experience it as pleasurable (a nice meal, our favourite music) or as unpleasurable (fingernails grating against a blackboard, root canal treatment). Additionally, Epicurus equates pleasure with goodness and pain with badness. This approach can be broadly labelled as hedonism (from the Greek *hedon*, meaning pleasure). More accurately, to borrow Fred Feldman's definition, hedonism is the view that a life is better for the one who lives it if it contains a more favourable balance of pleasure over pain. The way to determine how good or bad something is, is by measuring how much pleasure or pain it causes someone. This view has, of course, different versions.

Some are more, well, hedonistic (in the everyday meaning of the word). According to these views, the ultimate good is pleasure and the only thing that matters is obtaining it. We normally associate this type of hedonism with excess, putting one's pleasure above everything else, overindulgence and intemperate behaviour. We could object to this view on at least three counts. We could say that this kind of hedonism is objectionable because it is harmful (to ourselves or to others), because it is ultimately irrational (it is self-refuting, as I explain below) or because we see certain kinds of behaviour as immoral, for example, putting your own pleasure before others'.

Other types of hedonism, like Epicurus', are much more rational and differ considerably from the above view. In fact, Epicurus explicitly opposes "vulgar" hedonism. He writes: "when we say that pleasure is the goal we do not mean the pleasures of the profligate or the pleasures of consumption, as some believe ... but rather that lack of pain in the body and disturbance in the soul" (1994: 30–31). According to Epicurus, what matters is not only pleasure but more importantly, the avoidance of pain. He thought it natural for human beings (and other animals) to try to maximize pleasure and minimize pain. The absence of pain is already, in itself, a great pleasure. But more pleasure, once the absence of pain has been achieved, is not more valuable. It does not add anything over and above the pleasure already obtained by the absence of pain. (If this sounds odd, think of the disappearance of a toothache after taking painkillers. Epicurus is pointing out that the neutral state of painlessness, both physical and mental, is not to be taken for granted.) It merely makes it more varied.

For example, if someone is very thirsty she can derive great pleasure from quenching her thirst. Of course one can quench one's thirst with tap water or with champagne. Epicurus would say that the correct thing to do is to choose water over champagne. Why? Because the main aim is to rid ourselves of the painful sensation of being thirsty. If we drink water, we have created a very pleasurable feeling of not being thirsty anymore, using relatively modest and accessible means. If we drink champagne and develop a taste for it, champagne will become the condition of feeling pleasure. And if we no longer have champagne, we will still need it to feel pleasure and so have failed in our quest for leading a good, pleasure-filled life.

We can now begin to see why simple sensory hedonism (the view that pleasure is simply sensory pleasure) is self-defeating. This is why we said earlier that "vulgar" hedonism is irrational: because ultimately it leads to dissatisfaction, being unable to fulfil all our desires all the time. It is not enough to indulge oneself continuously by pursuing pleasure; rather, pleasures have to be selected that are necessary and natural, rather than unnecessary and unnatural. To continue our champagne example, if you derived a lot of pleasure from very expensive champagne and drank it every day that would be fine. But what would happen if one day you could no longer afford this champagne or if the winery closed down? You would still desire the champagne but would be unable to satisfy this desire. The result would be unhappiness.

In order to avoid this problem, Epicurus suggested only fulfilling desires that are necessary (such as quenching your thirst) and natural (quenching your thirst with water rather than champagne). If we ensure that our desires are natural and necessary, we will not become trapped in a rampant hedonism that sacrifices everything else for the sake of pleasure. Nor would we risk cultivating pleasures that may at some stage be unattainable. If we limit the desires we have, we become more autonomous and less dependent on external factors, such as the presence of champagne. And the less dependent we are on external, transient factors like champagne, the more tranquil we can be. If we are autonomous and rely on a minimum of material goods for our well-being, our tranquillity cannot easily be taken away from us or be affected by external change of fortune. And ultimately, it is this tranquillity that Epicurus is after.

So Epicurus endorses hedonism, but of a more sophisticated and thoughtful strand than the one wrongly associated with his name. His version of hedonism is well thought out and based on a rational, surprisingly modern view of the human being. According to Epicurus, we are material creatures influenced by material conditions, such as heat and cold, hunger and satiation and so on. Because our well-being is so intimately connected with our sensations and feelings, we must ensure that we experience as much pleasure and as little pain as possible. But this is not achieved merely by indulging our bodies. It is also achieved by scrutinizing different kinds of pleasure and cultivating some (necessary and natural ones) while culling others (unnatural and unnecessary ones).

In addition, Epicurus divides pleasures into mental and bodily and into static and kinetic. Static pleasure simply is the absence of pain. In the mental realm static pleasure is experienced as *ataraxia*, which is translated as tranquillity or freedom from mental disturbance. In the bodily realm pleasure would express itself as painlessness or *aponia*. Kinetic pleasure is the stimulation needed in order to arrive at static pleasure, or the absence of pain or mental anguish. So drinking water to quench one's thirst is a kinetic pleasure required to achieve satiety, which is a static pleasure.

From our previous discussion of death and with this overview of Epicurean hedonism in place, we can now see that one of the principal achievements of Epicurus' ideas was that they enable us to get rid of irrational and erroneous beliefs. This, in turn, is a substantial source of two kinds of pleasure: kinetic mental pleasure arising from solving a philosophical riddle (finding an answer to the question "Should I fear death?") and static mental pleasure arising from the dissolution of irrational beliefs that cause us mental agitation. This will bring us one step closer to achieving *ataraxia*, mental tranquillity.

Later thinkers continued Epicurus' line of thought and argued that pleasures do not have to be merely sensory. Pleasures can also be found in our attitudes and thoughts. We could take pleasure in thinking about a problem in philosophy, reading a good novel or thinking about world peace. None of these sources of pleasure is sensory. Attitudinal hedonism is the theory that makes room for these kinds of pleasures. On this view, the source of pleasure and pain can include attitudes we have towards different things. Attitudinal hedonists think that hedonism is not only about sensory pleasures, which are just part of the story. Their version of hedonism is more complex:

it consists of natural and necessary sensory pleasures, of having pleasurable attitudes and thoughts and, additionally, of having a pleasurable attitude towards one's life as a whole.

SHOULD WE FEAR DEATH?

We can now return to the fear of death and see how on the Epicurean view ceasing to fear death is a condition of the good life. For Epicurus (and other ancient thinkers) the fear of death was a major obstacle for achieving tranquillity and therefore had to be thoroughly tackled. As was said earlier, Epicurus held rational materialist views. He thought that the body and the soul were made out of matter (atoms) and that the soul perishes with the body. Therefore, there is no possibility of experiencing either pain or pleasure (mental or physical) after death. This, for Epicurus, is good news. It is a positive fact because it means that fears and threats of eternal damnation and of posthumous torture are unfounded and the source of needless suffering. Because he does not believe in the soul's survival after the death of the body, Epicurus is fully entitled to claim that there is nothing to fear in death, because death is pure non-existence, a state that cannot be compared to any other state (whether pleasurable or painful) or to any imagined existence in the grave.

Lucretius (*c.*95–52 BCE) was a follower of Epicurus (although he was born nearly 200 years after Epicurus' death) who wrote a long poem called *De Rerum Natura* (On the nature of the universe). In this poem he argues that the main reason people reject Epicurus' rational views about the fear of death is a failure of imagination. We are incapable of imagining our own non-existence. We can only imagine ourselves lying in a coffin, or watching our funeral and so on. But this, claims Lucretius, is precisely not what Epicurus is asking us to think about:

> When a living man confronts the thought that after death his body will be mauled by birds and beasts of prey, he is filled with self-pity. He does not banish himself from the scene nor distinguish sharply enough between himself and that abandoned carcass. He visualises that object as himself and infects it with his own feelings as an onlooker. That is why he is aggrieved at having been created mortal. He does not see that in real death there will be no other self alive to mourn his own decease – no other self standing by to flinch at the agony he suffers lying there being mangled, or indeed being cremated. (Book 3, ll.879–89)

Lucretius argues that it is our failure to imagine our non-existence that makes people resist Epicurus' arguments that "death is nothing to us". This seems to be a key obstacle for adopting Epicurus' cure for the fear of death. Some put the point more strongly, and claim that imagining our non-existence is not only difficult, but impossible. Anything I imagine has me in it as the imagining entity, the one doing the imagining. Therefore I cannot imagine any scene without putting myself as the viewer of that scene. Even if I remove myself from it, I am still present as a viewer. Whenever we try to imagine our non-existence we imagine ourselves witnessing some sort of state of affairs (it doesn't matter if it is pleasurable or not). At the very second we try to imagine this, we already fail, because someone must be there doing the imagining (this claim is analogous to Descartes' *cogito* argument).

It is impossible for us to imagine anything without having ourselves there to witness whatever it is that we are imagining. Therefore we are incapable of understanding our own non-existence and, as a result, incapable of embracing Epicurus' argument about why we should not fear death.

This seems to be a strong psychological claim about a general failure of imagination characteristic of most people, when asked to imagine their non-existence. But it may be possible to sustain the Epicurean argument by asking people not to *imagine* their non-existence but to *think* about it, to try to conceive it. Lucretius' vivid description of a corpse being devoured by beasts of prey could be replaced by a more rational request to ask people to conceive their non-existence. This may be more in line with Epicurus' rational approach to death as well as an easier task.

Some criticize Epicurus and ask whether by extinguishing our fear of death he has also extinguished our desire and love for life. This seems to be a weak criticism. From the fact that someone does not fear their death (because they rationally understand that there is nothing to fear in non-existence) one cannot deduce that they do not love life. It is perfectly compatible to believe in posthumous non-existence and therefore not fear death, while enjoying and cherishing the pleasures and experiences life affords us.

Another criticism compares non-existence with existence and concludes that the latter is better than the former. In other words, if I die I will miss out on all the fun things I could have done had I been alive. This view is called "deprivation theory". It argues that death is bad because it deprives us of possible goods that may have come our way had we still been alive. One philosopher who supports this view is Thomas Nagel. Nagel argues that "if death is an evil, it is the loss of life, rather than the state of being dead or nonexistent, or unconscious, that is objectionable" (1993: 63). So death is an evil and it is reasonable to react to it with fear because it denies us goods we could have had, had we been alive.

This may seem like a strong argument but the Epicureans have a good response. They can reply that deprivation theorists are not comparing like with like. We can compare playing the piano to eating ice cream and argue about which one is better. But we cannot compare our state of existence, whether good or bad, to our non-existence. Existence will have some quality to it: it will be good, bad, neutral or a mix of the three. Non-existence, on the other hand, has no quality; there is nothing it is *like* to be dead, so there is nothing to compare existence with. The two are simply incommensurable. Of course, one can say that it is not the presently dead person who is being deprived but the person while she was alive. But to this the Epicureans can say that the person and the harm can never coexist. If I am alive, I am not harmed or deprived by death. If I am dead, I can no longer be deprived of anything.

The Epicureans can also say that by Nagel's own lights we should *always* fear our death, even if we live to be 120. We should always fear it because on Nagel's view, whenever death strikes, even if I am very old, it always deprives me of living one more day. Similarly, we should always regret missing future events that we will not be able to witness because of our death, even if they are beyond the reach of any human being of our generation (say, celebrating the third millennium). This life full of regret does not seem an appealing way to live.

Deprivation theory has additional problems. I shall explain what I think is wrong with Nagel's view not only in order to defend Epicurus, but also to claim that his view does not take into account ill people. Moreover, focusing on the future, as many of Epicurus' critics do (for example, by arguing that we should want to live as long as possible), is not necessarily conducive to well-being.

The first problem of deprivation theory is that it is not clear that life itself, once we remove the positive and negative elements in it, is "emphatically positive", as Nagel describes it. Why is existence good in and of itself? Nagel says this: "there are elements that, if added to one's experience, make life better; there are other elements that, if added to one's experience, make life worse. But what remains when these are set aside is not merely *neutral*: it is emphatically positive" (1993: 62). But this begs the question. Nagel does not explain where life's positive value comes from. Nor does he answer the question: what makes life (with both good and bad elements removed) emphatically positive?

We can imagine a horrible life, full of suffering, pain and anguish, which would lead the person living it to suicide. The rationale behind the suicide would be that this kind of life is not worth living, that having no experiences at all is preferable to having these terrible experiences. Similarly, we can imagine a good life, full of pleasure and enjoyment. We can imagine that the person, when alive, does not wish to die. But what do we say of Nagel's pure life? What remains when good and bad are set aside? In this case there seems to be no reason to grant life any value other than a neutral one.

Life is a set of potential experiences, potential actions and potential good and bad events. In itself it bears no value whatsoever. It is the condition of possibility of having experiences, the space within which experiences take place. But as we have seen, these experiences can be either good or bad. There is no *a priori* guarantee that one's life will contain more good than bad, or more pleasure than pain. Life itself, the mere fact of existence – prior to knowing if it contains good or bad experiences – is neutral. The case for ill people is more acute. If illness is progressive and there is less and less of a chance that the ill person will experience good things, and more of a chance that she will experience bad things, why should she be burdened with the (unsupported) demand to want more of it? I can certainly imagine a day when I will think that my life is (for me) no longer worth living. If I suffer pain and no longer experience much pleasure in anything, why should my existence in itself add positive value to my life?

This takes us back to a previous problem, that of not being able to compare existence and non-existence. Again, those criticizing Epicurus' view seem to say that if someone dies today, they are deprived of the enjoyment of eating ice cream tomorrow. But in what sense are they deprived? As we said previously, there is no one to be deprived of anything once the person dies. Moreover, if we take this proposal seriously then we are constantly being deprived in a myriad of ways. I may have been deprived of a great party my parents threw a week before I was born. I may have been deprived of not experiencing things that happen in other places. If I live in the UK, I may have been deprived of having fun in Barbados, Cuba and China. Perhaps I am also deprived of not having as many friends as I could have had if I spent more time socializing, of not being as strong as I could have been if I worked out more, and so on.

The things of which I may be deprived are not unrealistic. I am not claiming that I am deprived of wearing the crown jewels or playing for Manchester United (examples proposed by

Michael Brady). But I can reasonably claim that things could have been better for me in at least some respects, so I am deprived of having a better life. Nearly all of us are deprived in the sense that our lives could have gone better, at least in some respect, than they actually have. And so, deprivation theory's objection to Epicurus makes less sense once we see that there are no real limits to what one may *reasonably* claim to have been deprived of.

Additionally, in situations of uncertainty about the future, life may cease to seem emphatically positive. In fact, life becomes a burden and maintaining a positive outlook on it becomes increasingly difficult. Perhaps those who reject Epicurus are thinking and writing from the point of view of a relatively young, fairly healthy person. But this point of view – implicitly taken to be the average one – is at odds with an ill person's point of view. Life seems emphatically good to those who think that it holds good things in store for them. But if this implicit assumption is removed, what basis is there to think of life itself as emphatically good? It makes more sense to think of life as emphatically *neutral*: if filled with positive content, it is good. If filled with negative content, it is bad. In itself it bears no value.

Let us turn to the second claim deprivation theorists make against Epicurus. Nagel says that life is a good, and "like most goods, this can be multiplied by time: more is better than less" (1993: 62). This, too, seems counterintuitive. In what sense is more better than less? It makes sense to say this of some pleasurable experiences, although by no means all. We normally think that it is better to have more money, more holidays, more friends and so on. But think of the simple bell curve that seems to describe many pleasurable experiences. When we go into a hot shower (to use an example from Fred Feldman) we get pleasure out of its pleasant warmth. If we make it hotter, it may become even more pleasurable for a while, but then it will become unpleasurable. If we continue to make the shower even hotter, it will become downright painful. A similar bell curve applies to eating chocolate or drinking champagne. There is a limit on how much pleasure we can get out of most things and at some point more stops being better than less.

But even if we grant this point (that more is better than less), there are further problems. Let's say that we manage to overcome this problem by getting the proportions right: say that we eat just the right amount of a lovely breakfast. It is nicer to have a lovely breakfast every morning, rather than just once a week. But if I have a lovely breakfast every morning does it matter whether this pleasant experience repeats itself five thousand or ten thousand times? It only matters if one day I cease to have lovely breakfasts but continue to exist. Then I will feel the lack of the lovely breakfast. But if I cease to exist (and of course cannot have breakfasts, lovely or otherwise, anymore), in what sense am I missing out on anything here? I am not missing out on anything because lovely breakfasts – or anything else, for that matter – can make no difference to me whatsoever.

Nagel is arguing for an objective sense of "better" and "worse", as if someone were standing above and adding up the good and bad experiences I had. From the objective observer's point of view, having ten thousand rather than five thousand breakfasts may seem like a good thing. But if we consider this point of view carefully, it does not make much sense. If someone is viewing the number of lovely breakfasts *objectively*, it doesn't matter to the observer whether I had few or many. The objective point of view is impartial. The observer is taking stock of my experiences, not having them.

If I am viewing the number of lovely breakfasts *subjectively*, then what is important is that I have them as long as I am there to enjoy them. From the subjective point of view, why should I care that I will no longer have lovely breakfasts when I cease to exist, when I know that I will not be there to miss the breakfast? To care about this seems to be incoherent: in what sense does it make any difference to me? Moreover, to care seems to cast a shadow over the present, when I am still here and still having the breakfasts and therefore seems to lessen the good I am experiencing while existing.

As we saw, more is not better than less in many cases. It certainly doesn't seem to be the case when we think about life. Compare three cases: in the first, the person dies at twenty-five, having lived a perfect, blissful, immaculately happy life. In the second case, the person dies at fifty, having lived a mixed life, with a balance between good and bad experiences. In the third case, the person lives to be a hundred, but his life is a tormented, unhappy life. It does not seem too difficult to choose the first person's life over the third person's, because not having lived to be a hundred is not a loss to the person (although it may be a loss to the person's family and friends), whereas one hundred years of suffering are not a particularly attractive option.

What about the second case, the life that is a balance between good and bad? Even in this case, it seems to me that the first life is preferable because it contains uninterrupted goodness. It doesn't make sense to want more of something, unless that thing is good (or at least it is probable that it will be good). We don't just want more life, we want more good life. In some cases of illness, the odds seem to be stacked against the person. There is a promise (and often a reality) of suffering of various kinds. So we should not be forced to desire more life, if that life does not seem good to us.

FEARING DEATH IN ILLNESS

We now know what Epicurus and Heidegger thought about the fear of death and the good life. We can examine which view is more amenable to a life that is already temporally and physically compromised and limited by illness. For ill people, there is a great need for both points of view. There is a need to come to terms with one's finitude and to appreciate fully the passing of time. There is also a need to overcome the fear of death that Epicurus takes to be so irrational. And ultimately, the two positions may not be as far from each other as we initially thought. Perhaps in order to be authentic – to live well – we must overcome our fear of death. And perhaps in order to overcome our irrational fear of death we must learn to see our lives as a finite whole.

I do not wish to belittle the profound differences between Epicurus and Heidegger, but there seem to be some affinities between the views. Ultimately Epicurus suggests that we cease thinking about death. Once we figure out that death is simply non-existence it no longer casts a shadow on our life. Heidegger, on the other hand, asks us to *anticipate* our death, to foresee its inevitability by choosing to live authentically. But Heidegger also tells us that anticipating death does not mean dwelling on it, obsessing about it or living in constant anxiety of it. Anticipating death means living life as finite and understanding our finite structure more fully. Heidegger uses the term "to understand" here in a practical sense – to make choices, act, respond to situations – in short, to *live* as finite. Epicurus seems to recommend something that is not too dissimilar. He seems to recommend making peace with our finitude by understanding that non-existence is not something to fear and then living accordingly.

There are other points of similarity. Both Heidegger and Epicurus emphasize the importance of understanding our finitude; for both of them the notion of an afterlife is simply a way of obscuring the serious question of mortality each individual must face. Additionally, both focus on mental anguish (which Heidegger calls anxiety) in relation to death. While Epicurus wants to cure the soul from anguish and lead it to tranquillity, Heidegger sees anxiety about death as a springboard to authenticity. We must pass through anxiety in our quest for the good life, or the authentic life, because anxiety indicates that we understand the significance of death for life.

There is another reason I chose to bring the ideas of Epicurus and Heidegger to bear on illness. For both, the question of how to live life while knowing that it is finite is not a theoretical question. Both regard death not only as a philosophical issue, but as an existential concern that we must address in order to live a good life. Although they have different ideas about death, both Epicurus and Heidegger work towards unifying the first-person existential perspective with the third-person objective perspective of philosophy. The dialectic between the personal and the philosophical I spoke about in the Introduction is also very apparent for these two philosophers. I shall say more about the relevance of philosophy to our personal lives in the next chapter.

I now turn to the special requirements that become pertinent when we think about the prominence of death for gravely ill people. It is one thing when death is a far away possibility that will take place in the distant future. It is another thing altogether when death is a conceivable possibility in the near future, as it is for those whose prognosis is poor. The sense of uncertainty, the fear of leaving loved ones behind, of leaving incomplete projects midway, of not saying the important things, the worry about what will happen to the ones I am now able to love and protect – all these are no longer abstract, hypothetical issues but pressing and immediate concerns. The perennial question – What do I do in the face of adverse circumstances over which I have no control? – becomes urgent and demands to be addressed here and now.

One way to respond to these practical problems is to investigate the relationship between our conceptions of time and death and our ideas about what constitutes a good life. But this cannot be done merely by taking general ideas about these concepts and applying them to illness. It is not enough to say that whatever we think about death or time will apply to the case of illness, but will simply be more imminent. Rather, in the case of illness, these questions – What should I do with a limited amount of time? How do I live when death is impending? – require a *practical* response. They need to be addressed not only theoretically, on paper, but in action, in deeds.

The theoretical discussion of death and temporal uncertainty has little in common with their actual lived experience. Reading about death and illness is nothing like receiving a dire medical prognosis. The cramped doctor's office, the tightening chest, the cold panic washing over you; these are so visceral, so traumatic, so real. It is this gap between the two – the theoretical contemplation and the lived experience – I aim to bridge here. I aim to bridge this gap by providing a dialectical account of a philosophical theory that has become a lived experience and is then rethought in light of experience.

One of the courses I teach at university is a course on death. When I teach my students, most of them in their early twenties, about death, I often feel that for them it is a fascinating intellectual exercise rather than an honest survey of their own lives. I hide my illness behind their youth and inexperience (most of them are lucky to be ignorant of the personal impact of illness). For these young adults, such questions are deeply academic: fascinating, but distant. These are very young people, at the start of their adult lives, for the most part untouched by death and tragedy. We spend many hours reading papers on death and discussing the various philosophical issues that arise. Would immortality be boring? Is death a harm to the person dying? Does finitude detract from life's value?

I also feel like a fraud. Here I am, calmly discussing these questions and intellectual puzzles while inside me a multiple-injury train crash has taken place. A global disaster has wrecked my life, progressing at a cruel and barely comprehensible rate. But when I teach, I distance myself from my illness. For the most part, my students have no clue that the issues we are debating in this hypothetical, rational way are the same issues that dictate my life's expectations, possibilities and reformed narrative. That my illness has turned my life upside down, inside out, back to front. That a chaos presenting itself as order, a well of grief presenting itself as cheerful everydayness, a scream silenced by my good upbringing, are packaged neatly within the woman teaching them philosophy.

This fraud, this lie, is something I have been forced, or advised, to live when I teach, when I present my work, when I write papers for academic journals. It has happened that in a personal chat after a talk I occasionally tell people that what drew me to write about illness was my personal experience. "Well done," they say, "I never would have guessed." I have been commended on my "professionalism" because I removed myself from the more abstract issue. But in fact what I do is dress the most personal, the most pressing issues for me, in academic robes.

I do not mean to belittle the importance of reflection and philosophical thought on personal issues, such as illness and death. On the contrary, philosophy has been my strongest ally in coming to terms with life as an ill person. There is real pleasure and tangible reward in applying philosophy to one's own life. But in order to do so, we must let go of the idea that philosophy is abstract, objective and academic. We must return to philosophy with lived experience to fuel and feed it, to motivate questions and enquiry. Philosophy is also a real-life tool, a form of therapy that uses reason to combat fears, in the way we have seen Epicurus do. It is an ability not just to think but to apply thoughts, concepts, critique, to one's life. In this sense philosophy is not only a quest for a good life, but also a concrete path to achieving it.

But in order to apply philosophy to lived experience the gap between the two must be overcome. I said earlier that philosophy, at least in the way it is practised today in Western universities, is seen as objective and impersonal. This is at odds with the very personal nature of the problems we have been discussing. My attempt here is to try and bring these two antagonistic forms of experience and expression into conversation with each other. It is an attempt to bring together philosophical insights and personal experience and to create dialogue between the two using phenomenology. The point of this dialogue is to show how an existential problem like illness can be addressed by philosophy. In the final chapter I take this idea further and develop the notion of philosophical therapy and show how such therapy can modify our notion of time.

FIVE
Living in the present

What would you do if you were told you have a year to live? A month? A day? You would probably have very different plans for each scenario. Our diverse attitudes to life stem, in part, from our different estimates of how much time we have. Some people think that life is very long; that seven or eight decades of living are more than enough. Others believe that they would be happy if they could become immortal. Still others think that immortality would be intolerably boring. Each one of us has her own notions of time, mortality and the good life.

Moreover, the three are linked in ways that illuminate illness. As we saw in the previous chapter, you cannot have a good life if you are constantly plagued by fear of death. Other fears of what may happen in the future could trouble us, preventing us from living well in the present. Memories of a difficult past may also interfere with our present well-being. In this chapter I will show how our ideas about time and well-being are related and how certain philosophical views may lead us to a happier life, even if it is lived in the shadow of illness.

Let us begin with time. Is the average life span in the West, say seventy-five years, long or short? Do we have plenty of time or too little of it? Our experience of time depends on what we want to achieve and how much we are enjoying ourselves while doing it. If I wanted to dig a tunnel to the other side of the earth using a spoon, an average life span would be too short for achieving this aim. If I wanted simply to enjoy myself, have a good time, eat nice food and so on, it wouldn't matter very much if I did these things for twenty, fifty or a hundred years (assuming that the level of enjoyment is the same throughout my adult life). So life is not long or short; it is long or short relative to what we want to do. Some projects require many decades, others are short-lived.

Similarly, our consciousness of time influences our sense of whether a given period is long or short. Waiting notoriously slows down time for us, while enjoyment makes it zoom past. As a child I remember time slowing down noticeably during the last period at school, being tortuously slow on the count-down to my birthday or a school trip, but the summer months always flew past too quickly.

So there is no fact of the matter about whether the average life span is long or short. But there is an average in our privileged Western world, which tells us that we can expect to live roughly 75–85 years, give or take a few. And we adjust our plans and expectations accordingly. We postpone some things, while rushing to do others. We spend decades in education and training. We go through childhood, adolescence and youth feeling that time is plentiful. We coast through our twenties and thirties with the implicit sense that there is plenty of time. That time is an abundant resource we can squander, because life is so long and there is still so much of it. Being old is something that will happen later. Dying is something that will happen even later than that. No need to think about it now.

Common thought patterns that accompany this attitude are "someday" and "later". Someday I will learn to play the piano; someday I will travel to Africa. Postponement, procrastination and delay are prominent elements in our lives. We usually feel that there will be time – later – for doing the things we do not want to do now. This sense of an unlimited supply of "later" is mirrored by a corresponding sense of regret when we look back at missed opportunities, paths not taken. But time always moves forwards and rarely do we have occasion to correct past mistakes or pursue missed opportunities.

We think about time when designing our plans and projects. We normally have a good idea what we'll be doing next week, some idea what we'd like to do next summer and a vague five-year plan. We also have, possibly more implicitly, a more general life plan: where we would like to be in ten, twenty or thirty years. These plans are based on our reasonable expectation that we will be alive for the foreseeable future and that our loved ones will be too. We are not fans of drama when it comes to our life. We mostly just want things to stay the same: not to age, not to lose our parents, not to see our children leave home.

All this changes when you are ill. Life ceases to be a long, gently flowing river. The future no longer contains the vague promise of many more decades. Death is no longer an abstract, remote notion. The soft-focus lens is replaced by a sharp magnifying glass through which terminal stages of illness can be viewed in nauseating detail. The future curls in on itself and at once becomes both exposed and radically curtailed. It has a clear endpoint.

Several times when I told people about my illness they asked: "So how long have you got?" The question always left me gasping for air. After overcoming my horror at the casualness with which the question was asked, I wondered why they wanted to know. I think it was because they wanted to place my story within a temporal framework. They wanted to know how bad it really is. And "how long I have" is one way of getting a grip on how bad it is.

Time did change for me. I began to take it much more seriously. I began to make a point of enjoying things thoroughly: memorizing sensations, views, moments. Partly in preparation for days to come in which I may not be able to leave the house or my bed, but also in order to feel that I have taken the time to really sense, really experience pleasurable things. I wanted to feel that I am living life to the full in the present. That I *am* now.

By focusing on the present I learned to discount the future, while it seemed to me that so many of my friends were doing the opposite. They seemed to be always waiting for something to happen: the promotion, the birth, the trip. I had nothing to wait for but bad news. I no longer cared about promotion, I knew I could not have children and I could not travel. The only news that routinely came my way was bad news: a further deterioration, worrying cysts in the pelvis, more pain and disability. My future held no promise but the promise of decline and a rapidly shrinking world.

Because I am exposed to so much uncertainty, I had to find a way not to worry about it too much. I am at risk of many complications. My lungs could collapse at any time, rendering them useless. Lymphatic fluid accumulating around the lungs could send me to the hospital. Kidney tumours could cause internal bleeding. Lymphatic blockages could cause pain and swelling in the abdomen ... the list is long and scary.

In the first days after my diagnosis I was terrified and walked delicately, sensing danger everywhere. After a while, I stopped thinking about it. When chest pains came, I accepted them. When my lymphatics did dilate causing continuous pain, I accepted that too. Yes, Really Bad Things could happen to me at any minute. But not now. And now is where reality is: liquid time solidified into a crystal drop of Now. I grasped that drop with both hands, clutching, savouring, enjoying. Now became the place for me, too, to be.

The fragility, but also the preciousness, of the present became a fundamental building-block of my experiences. I learned to dwell in the present and stop the train of thoughts about the future, about all the scary painful things that could happen, about needing a lung transplant. Things that seemed surreal became my daily bread. Using oxygen, having a life-threatening illness, suffering severe breathlessness at every twist and turn of my daily routine; these are with me when I first open my eyes each morning and are the last thought of the day. I struggle not to let them fill every thought in between.

Most nights I wake thinking about it. In the beginning I was terrified. I would wake up my husband; tell him my thoughts and fears. Talking about it didn't help, but surrendering to the thoughts did. Slowly I learned to succumb. I now lie there, thinking how badly things turned out for me. How scared I am. How little control over it I have. And eventually my mind runs out of steam, the thoughts trickle through their course and I return to sleep. Don't fight it, I learned. The thoughts will come and go. Battling against your own mind is a bad strategy in these situations. It is better to let go. And so I wake up and lie quietly in the dark, no longer frightened. The thoughts swirl in my mind. I let them pour through my consciousness and then vacate that space, making room for sleep. We learn to live with the strangest things.

More practically, some things did not seem worth investing in any more. I no longer save money. I ceased to make extra payments into my pension fund. When I want to do something, I no longer hesitate or feel guilty about spending time or money. There is no point saving for a rainy day. For me it is a hurricane every day of the year. I indulge in many things I would not have permitted myself when I was healthy. I care much less what people think of me. It is liberating to live in the now. It is liberating to be freed from having to plan, to make a future, to strategize.

I found to my surprise that I experienced amplified enthusiasm and joy, echoing against the narrow confines of the present. All my energy and happiness are funnelled into the *now*, into today: how nice it feels to be *here*, in the sun, having a massage, listening to beautiful music, laughing until I am dizzy, sitting by a warm fire, experiencing friendship, love, sunshine, the lazy sensation of waking up after a deep sleep, the sharp authority of beauty.

I stopped looking ahead. I no longer think of my life as culminating with anything: having children, promotion, a successful grant application. Things that seemed like tremendous milestones are now optional. Each may or may not happen, but it will not have the significance I once attributed to it. I don't make plans for next year or even six months ahead. When people say "I'm planning a trip around the world in two years, after my promotion", I shudder internally at their hubris. My projects are now modest: a paper, a short book, a few conferences in the spring, a brief trip to Paris next month. Even they are in the shadow of substantial conditions: if I am able to travel to France by train; if I am not on the waiting list for a transplant.

In the same way that my geographical horizon shrank because I was no longer able to fly, my temporal horizon became truncated. And although I accept invitations for talks and trips months ahead, I always think: God knows what will happen by then. The long gentle river has become a tempestuous series of rapids.

PHILOSOPHY AS THERAPY

In this book we explored the lived experience of illness. We considered how life changes in illness, why the naturalistic view of illness is insufficient, how the social world of the ill person is transformed and how health within illness is possible. We then turned to the fear of death and presented different suggestions on how to cope with this fear. Our final task is to bring these ideas to bear on a fundamental question for an ill person: how should I live *now*? What kind of life should I lead in the present? The two main conclusions so far – that illness should be understood from a phenomenological perspective and that a good life is possible within the constraints of illness – have to be given a concrete meaning. We need to specify what a good life with illness is. My proposal is that one way of living well with illness is by living in the present.

Learning to live in the present with illness is learning to be happy now, regardless of threats to our future. It is learning to confine memories of past abilities and fears of the future so that they do not invade the present. It is learning to delimit them, stop them from shadowing the present. By changing our attitude to time, we can change the quality of the present, how much we enjoy living *now*.

In using philosophy to figure out how we can live well in the present I hope also to show its important role in providing guidance in everyday life. In presenting this view, I continue an ancient philosophical tradition of seeing philosophy as a practical aid to life. We will use philosophy (much as we have been doing throughout the book) as a tool to improve well-being. And there is nowhere better to start than with the ancient Greek philosophers, who saw philosophy as a primary aid in our quest for the good life.

In the previous chapter I chose to present the views of Epicurus and Heidegger because of an important commonality in their positions. For both thinkers the philosophical question of how to face death is at the same time personal, pressing and concrete. More generally, in ancient Greece philosophy had a practical use. It was not an abstract, theoretical activity, disengaged from everyday life. On the contrary, philosophy was the tool with which to analyse, criticize and ultimately improve everyday life.

This is obvious in the writings of Epicurus. He thought of philosophy as medicine for the soul, a form of therapy in words. For him, philosophy presents therapeutic arguments that ultimately help people see reality more clearly and have more accurate concepts with which to understand their lives. "Empty is the argument of the philosopher by which no human suffering is healed; for just as there is no benefit in medicine that does not drive out bodily diseases, so there is no benefit in philosophy if it does not drive out the suffering of the soul", he wrote (1994: 99, translation modified). As Martha Nussbaum says, we can use philosophy to understand the ways in which human lives are diseased and what they need. But this is only ever a prelude to then trying to heal those lives and give them what they need. "The whole point of philosophy is human flourishing", she writes (1994: 34).

So how can philosophy cast out the suffering of the soul? In what ways can it help the quest for a good life? Epicurus saw philosophy as an activity that secures the good life through the use of arguments and reasoning. On Epicurus' view, mental anguish is produced by false beliefs. Good arguments against these false beliefs can dissolve this anguish like a painkiller dissolves physical pain. As Nussbaum writes: "the central motivation for philosophising is the urgency of human suffering, and the goal of philosophy is human flourishing" (*ibid.*: 15). Philosophy is an art whose tools are arguments, but these arguments are defective if they do not bring about a change in the listeners' beliefs. On this view, philosophy is committed to action and change, as well as to rigour and logical validity.

The recognition that our beliefs are wrong is a first, but substantial, step towards ridding ourselves of false beliefs and erroneous judgements. We may falsely believe that money is the most important thing in life. If we take seriously Epicurus' arguments about the insignificance of external goods, we may be released from our erroneous belief that money is so important and cease to suffer because we do not have enough of it. Similarly, we may believe that good health is a condition for well-being and feel despair because of its absence. If we take on board the notion of health within illness we can be released from this belief and find happiness with partial health. These are examples of suffering created by false beliefs and the relief that can be brought about by their removal.

In previous chapters, we spent much time discussing the practical (but also philosophical) problems facing those afflicted with illness. The problems are manifold and take on a different form and intensity in each individual case. As we saw in Chapter 3, the plurality of the experiences of illness is striking. Nonetheless, there are general issues that ill people confront. These issues can be summarized in one question: how should I face adverse circumstances over which I have no control?

In this general form, the question becomes not only a common one for many ill people now, but also a question that has been with humanity for thousands of years, across cultures and religions. It is a difficult question to ask, because it forces us to admit that we have limited control over our lives. It is even more difficult to answer, because there isn't a really good answer to the question: what should I do when things go horribly wrong and no one can make them better?

This is where philosophy, or changing the way we think, is a potent tool. (There are other ways of achieving a change in patterns of thought, for example, cognitive therapy.) I cannot change reality; my illness is here to stay. But I can control some of the elements making up my life. I can, for example, control my thoughts (to some extent); I can control my reactions; I can cultivate the happy aspects of my life and I can say no to distressing thoughts and actions. I can choose what to do with the time I have and I can reject thoughts that cause me agony. I can learn to think clearly about my life, give meaning even to events beyond my control and modify my concepts of happiness, death, illness and time.

Throughout the book I have tried to show how this is possible. Taking on board the concept of health within illness and overcoming our fear of death are two ways in which we can change our attitudes towards the given facts of illness. Many ill people report that illness has given them a sense of pride in their resilience, a new understanding of their life and a deeper appreciation of the good things in it. Nietzsche seems to understand this regenerative capacity of illness when he writes about his own prolonged period of illness:

> It was as if I discovered life anew, myself included; I tasted all the good things, even the small ones, as no other could easily taste them – I turned my will to health, to *life*, into my philosophy ... the years when my vitality was at its lowest were when I *stopped* being a pessimist. (2004: 8)

Several philosophers think that focusing on living in the present and changing our views about time are fundamental to well-being. In the final section of the book we shall discuss this idea of living in the present and use the views of Hadot and Epicurus to tackle our question: how can I live well with illness?

"... ONLY THE PRESENT IS OUR HAPPINESS"

The French philosopher and ancient Greek scholar, Pierre Hadot (1922–2010), continues the Epicurean tradition in claiming that only the present is our happiness. In Chapter 4, we learned that for Epicurus the quantity and duration of pleasure are of no importance. Once we achieve a complete absence of mental and physical pain, additional pleasure does not increase the value of this state. Moreover, whether this complete absence of pain lasts a moment or a lifetime makes no difference to Epicurus, as perfect happiness has been achieved and prolonging it will not make it any more perfect than it already is. But in order to attain such a moment of perfect happiness, we must remove from it both grief about the past and anguish about the future. Learning how to live in the present and knowing the "healthiness of the moment" are the apex of Epicurean philosophy.

Hadot turns to Goethe for his views about antiquity's attitude towards time. According to Goethe, in ancient times the present instant was pregnant with meaning, full of significance and intensity. Each moment was lived in all its richness and was sufficient in itself. There was no need to turn to the past or to the future in order to experience the present fully. This ability to live in the moment has been lost, says Goethe; we no longer know how to live in the present. We take the future as our ideal and infuse it with meaning, hope and desire, all the while neglecting the present. We see the present as banal and trivial, as a portal to better, more interesting times. We no longer know how to act in and on the present. We have lost "that splendid feeling of the present" (quoted in Hadot 1995: 220).

The ancient ideal was to know both how to live in the present and how to utilize it. Knowing how to live in the present is having the ability to be content in the present instant, content with earthly existence. Knowing how to utilize the present is having the ability to recognize and seize the decisive instant, without regretting a past or desiring a future. Examples of seizing the moment are a dancer who knows the decisive moment for a movement, or a photographer who knows the exact moment to open the shutter, a moment that would capture a certain situation in the best way. Capturing each instant in its fullness, in its significance and uniqueness, is what true appreciation of the present brings. This ideal of being in the instant and realizing its immense significance is echoed in Rilke's ninth Duino Elegy:

> *Once* for each thing. Just once; no more. And we too,
> just once. And never again. But to have been
> this once, completely, even if only once:
> to have been at one with the earth, seems beyond
> undoing. (1982: 198–9)

The main lesson Goethe takes from the ancients is that of learning to free oneself from the past and the future in order to live in the present. Epicurus claimed that happiness can only be found in the present. That one instant of happiness is equivalent to an eternity of happiness, and that happiness can be found immediately, here and now. For him, the slightest moment of existence acquires infinite value if it is lived well.

So how can we learn to enjoy the pleasure of the present? As Hadot, Goethe and Epicurus tell us, we must learn to focus on the present without being distracted by memories of an unpleasant past or fears of the future. We must bring into the present moment pleasurable thoughts and actions. And we must realize that the quality of the moment does not depend on the quantity of time it lasts or the quantity of desires that it satisfies. The best moment is one that contains the least amount of worry and the most peace of mind. Having such moments seems achievable, even if one's life is burdened with illness. As the four-part cure tells us, what is good is easy to get.

The suffering of the soul discussed by Nussbaum is the source of our inability to be happy in the present. Desires, cravings and frustrations are obstacles to happiness in the present because they distract us from the fullness and beauty of the present and turn our minds to future demands or past failures. The purest, most intense pleasure can be obtained in the present. This pleasure does not depend on its duration and does not need to last long in order to be perfect. As Epicurus says, "finite time and infinite time bring us the same pleasure, if we measure its limits by reason" (quoted in Hadot 1995: 224). This may sound odd but makes sense if we use the following analogy, as Hadot does. A small circle is no less a circle than a big one. Similarly, perfect pleasure is no less pleasurable if it lasts an instant than if it lasts a lifetime. What matters is the quality of the pleasure at the moment it is experienced. The prolongation of pleasure does not change its essence.

The conclusion for Hadot is that pleasure is contained wholly within the present moment and we do not need to wait for anything else to obtain happiness. If we limit our desires in a reasonable way, by pursuing only natural and necessary desires, we can be happy right now. As the fourteenth Epicurean saying of the Vatican Collection tells us, "you are not in control of tomorrow and yet you delay your opportunity to rejoice. Life is ruined by delay and each and every one of us dies without enjoying leisure" (1994: 36). Seeing each day and each moment as an unexpected gift can bring us to the conversion Epicureans advocate. Ceasing to experience everyday moments as mundane, the present as tedious and the instant as insignificant is the way to achieve happiness now, immediately, in the present.

Epicurean happiness requires a shift in perspective. If we cease viewing happiness as a rare achievement and see it as a simple, attainable state, we can find it here and now. It is an attempt not to dismiss life's challenges and hardship, but to illuminate the attainability of happiness. We can now return to this book's epigraph and understand it in full: "The cry of the flesh: not to be hungry, not to be thirsty, not to be cold. For if someone has these things and is confident of having them in the future, he might contend even with Zeus for happiness" (*ibid*.: 38).

As we saw, dwelling in the past or placing one's hopes in the future both disrupt happiness in the present. Delimiting the present, or learning to live in the now, is another way of describing the Epicurean emphasis on the present. If we think about it, the past and the future do not belong to us. The past is already fixed and we cannot change it. The future has not yet happened, so we cannot command it. Both are not within our control.

Only the present and our perception and experience of it are within our control. If we shift our attention away from the past and the future to the present, we can concentrate on what we are doing and enjoy it. If we think about mental disturbance we can see that it is mostly anguish about future events or sorrow about past events. We are rarely anxious about the present moment. Concentrating consciousness on the present moment is a key to reducing anxiety about events over which we have little or no control.

How does this view apply to illness? I think that this privileging of the present can make a significant change to the experience of illness. It enables us to locate the source of mental anguish in the past (memories of things I could once do) and the future (fearing future suffering and death). We can then distance ourselves from the two by focusing on the present, and this can be a way to live well with illness. Focusing on present abilities, joys and experiences instead of worrying about a no-longer-existing past and a not-yet-existing future, is a way of avoiding some of the suffering caused by illness.

Illness can be a journey. Like some journeys, you do not always know where it will take you. This particular journey moved from personal experiences of illness to a philosophical exploration of their meaning. With the aid of phenomenology, it linked the personal-subjective and the philosophical-objective. It ends, or rather stops, here, in the middle. In the present. Where I am now. I do not know what the future will bring; no one does. But being here, now, is enough.

This does not mean that I have given up. On the contrary: I have become a patient activist. I spend a few hours each week working for LAM Action, trying to push forward research and drug trials and to support newly diagnosed patients. I was surprised to discover the extent to which patient advocacy and funding can support researchers' work. I help organize research meetings, promote tissue donation for research and write material for websites. I give talks to raise awareness and funds for research. I write and speak of LAM and of the chronic shortage of organs for transplantation on every possible occasion. My academic work now focuses on the application of phenomenology in medical and healthcare training. Doing this is another triumph over my situation, another way of saying no to despair. LAM is a disease that has few advocates because it is so rare. That is why I have become my own advocate, like many other women with LAM. We refuse to accept the ten year death sentence proclaimed by the diagnostic manual that first introduced me to LAM seven years ago.

And so I continue to ride my electric bike to work, go to yoga class and see friends and family. I continue to walk my dog, listen to music, write. I continue to live. Sometimes my illness makes life hard. It often takes up more time and space than I would like it to. But it has also given me an ability to be truly happy in the present, in being *here* and *now*.

LAM: facts and figures

Lymphangioleiomyomatosis (LAM) is a rare and often fatal lung disease that affects women with an onset in their child-bearing years. In LAM, a multi-system cancer-like disease, abnormal cells proliferate in the lungs, kidneys, abdomen, lymphatics and blood vessels. This leads to collapsed lungs, breathlessness, an accumulation of fluid in the chest and abdomen and occasionally bleeding from kidney tumours. The disease progresses at a variable rate but leads to respiratory failure in the majority of patients. There is no cure for LAM, although the drug sirolimus has been shown to prevent disease progression in some patients. There are an estimated 250,000 women worldwide suffering from LAM, including many who also have a genetic disease called tuberous sclerosis complex. A lung transplant is an option of last resort if a matched donor can be found, but there is a critical shortage of lungs for transplant. (Source: Professor Simon Johnson, Director of the National Centre for LAM, Nottingham, UK)

The following websites of LAM organizations provide information, help and research support:

www.lamaction.org
www.thelamfoundation.org

For the NHS Organ Donor Register see:

www.uktransplant.org.uk

Further reading

Currently there are three main approaches to the concepts of illness and disease: naturalism, normativism and the less-known phenomenological approach discussed in the book. Christopher Boorse is a well-known naturalist. His views are presented in "Health as a Theoretical Concept", *Philosophy of Science* 44(4) (1977), and in "On the Distinction between Disease and Illness", *Philosophy and Public Affairs* 5(1) (1977). For a critique of naturalism, see Rachel Cooper's paper, "Disease", *Studies in History and Philosophy of Biological and Biomedical Sciences* 33 (2002).

Normativism is discussed in Lennart Nordenfelt's *On the Nature of Health* (1987) and *Health, Science and Ordinary Language* (2001). An interesting discussion of both naturalism and normativism can be found in William Fulford's paper, "Praxis Makes Perfect: Illness as a Bridge between Biological Concepts of Disease and Social Conceptions of Health", *Theoretical Medicine* 14 (1993). A good collection edited by J. Humber and R. Almeder is *What Is Disease?* (1997).

The phenomenological approach is developed by Fredrik Svenaeus in *The Hermeneutics of Medicine and the Phenomenology of Health* (2000). Another author on the lived experience of illness is S. Kay Toombs. Especially interesting is "Illness and the Paradigm of Lived Body", *Theoretical Medicine* 9 (1988), and *The Meaning of Illness: A Phenomenological Account of the Different Perspectives of Physician and Patient* (1999). Toombs also edited the *Handbook of Phenomenology and Medicine* (2001).

On a phenomenological approach to disability see Toombs's "The Lived Experience of Disability", *Human Studies* 18 (1995). Also good, if somewhat dated, is Erving Goffman's *Stigma: Notes on the Management of Spoiled Identity* (1963). Ron Amundson's paper "Disability, Handicap and the Environment", *Journal of Social Philosophy* 23(1) (1992), explores disability as a relationship of body and environment. A more general philosophical reflection on space (which was discussed in Chapter 1) is Gaston Bachelard's *The Poetics of Space* (1964).

Although this was not discussed in the book, for those who are interested in a philosophical treatment of mental illness and the issues specific to it, see Rachel Cooper's *Classifying Madness: A Philosophical Examination of the Diagnostic and Statistical Manual of Mental Disorders* (2005) and *Psychiatry and Philosophy of Science* (2007). A recent survey paper is Matthew Broome's "Taxonomy and Ontology in Psychiatry: A Survey of Recent Literature", *Philosophy, Psychiatry, and Psychology* 13(4) (2006). For a sociological approach to illness see Simon J. Williams's *Medicine and the Body* (2003).

Good general introductions to phenomenology are Dermot Moran's *Introduction to Phenomenology* (2000) and David Cerbone's *Understanding Phenomenology* (2006). For introductory books on Heidegger see Stephen Mulhall's *Routledge Philosophy Guidebook to Heidegger and* Being and Time (1996) and Richard Polt's *Heidegger: An Introduction* (1998). For more on Heidegger's concept of death see Havi Carel, *Life and Death in Freud and Heidegger* (2006) and "Temporal Finitude and Finitude of Possibility: The Double Meaning of Death in *Being and Time*", *International Journal of Philosophical Studies* 15(4) (2007). Books on Merleau-Ponty include Eric Matthews's *The Philosophy of Merleau-Ponty* (2002), *Reading Merleau-Ponty* (edited by Thomas Baldwin, 2007) and Monika Langer's *Merleau Ponty's Phenomenology of Perception* (1989).

For more on Epicurus and on hedonism, I recommend Fred Feldman's *Pleasure and the Good Life* (2004) and James Warren's *Facing Death: Epicurus and His Critics* (2004). Martha Nussbaum's *The Therapy of Desire* (1994) discusses therapeutic arguments and Epicurean philosophy. An excellent collection on death is *The Metaphysics of Death* (edited by J. M. Fischer, 1993).

References

Amundson, R. 1992. "Disability, Handicap and the Environment". *Journal of Social Philosophy* 23(1): 105–18.

Bachelard, G. 1964. *The Poetics of Space*, E. Gilson (trans.). New York: Orion.

Baldwin, T. 2007 (ed.). *Reading Merleau-Ponty*. London: Routledge.

Buber, M. [1923] 1971. *I and Thou*. New York: Free Press.

Boorse, C. 1977a. "On the Distinction between Disease and Illness". *Philosophy and Public Affairs* 5(1): 49–68.

Boorse, C. 1977b. "Health as a Theoretical Concept". *Philosophy of Science* 44(4): 542–73.

Broome, M. "Taxonomy and Ontology in Psychiatry: A Survey of Recent Literature". *Philosophy, Psychiatry, and Psychology* 13(4): 303–20.

Carel, H. 2006. *Life and Death in Freud and Heidegger*. Amsterdam: Rodopi.

Carel, H. 2007a. "Can I be Ill and Happy?". *Philosophia* 35(2): 95–110.

Carel, H. 2007b. "Temporal Finitude and Finitude of Possibility: The Double Meaning of Death in *Being and Time*". *International Journal of Philosophical Studies* 15(4): 541–56.

Cerbone, D. 2006. *Understanding Phenomenology*. Stocksfield: Acumen.

Cooper, R. 2002. "Disease". *Studies in History and Philosophy of Biological and Biomedical Sciences* 33: 263–82.

Cooper, R. 2005. *Classifying Madness: A Philosophical Examination of the Diagnostic and Statistical Manual of Mental Disorders*. Berlin: Springer.

Cooper, R. 2007. *Psychiatry and Philosophy of Science*. Stocksfield: Acumen.

Cornwell, J. 1985. *Hard-Earned Lives: Accounts of Health and Illness from East London*. London: Routledge.

Epicurus. 1994. *The Epicurus Reader*, B. Inwood & L. P. Gerson (eds). Indianapolis, IN: Hackett.

Feldman, F. 2004. *Pleasure and the Good Life*. Oxford: Oxford University Press.

Fischer, J. M. (ed.) 1993. *The Metaphysics of Death*. Palo Alto, CA: Stanford University Press.

Freud S. [1916] 1985. "On Transience". In *The Penguin Freud Library*, vol. 14, 287–90. Harmondsworth: Penguin.

Fulford, W. 1993. "Praxis makes Perfect: Illness as a Bridge between Biological Concepts of Disease and Social Conceptions of Health". *Theoretical Medicine and Bioethics* 14(4): 305–20.

Goffman, E. 1963. *Stigma: Notes on the Management of Spoiled Identity*. New York: Simon & Schuster.

Hadot, P. 1995. "Only the Present is Our Happiness: The Value of the Present Instant in Goethe and in Ancient Philosophy". In his *Philosophy as a Way of Life*, 217–37. Oxford: Blackwell.

Heidegger, M. [1927] 1962. *Being and Time*, J. Macqarrie & E. Robinson (trans.). Oxford: Blackwell.

Heidegger, M. 1992. *History of the Concept of Time*. Bloomington, IN: Indiana University Press.

Heidegger, M. 1995. *The Fundamental Concepts of Metaphysics*. Bloomington, IN: Indiana University Press.

Humber, J. & R. Almeder (eds) 1997. *What is Disease?* Totowa, NJ: Humana Press.

Langer, M. 1989. *Merleau-Ponty's Phenomenology of Perception*. Basingstoke: Macmillan.

Lindsey, E. 1996. "Health within Illness: Experiences of Chronically Ill/Disabled People". *Journal of Advanced Nursing* 24: 465–72.

Lucretius 1994. *On the Nature of the Universe*, R. E. Latham (trans.). Harmondsworth: Penguin.

Matthews, E. 2002. *The Philosophy of Merleau-Ponty*. Stocksfield: Acumen.

Merleau-Ponty, M. [1945] 1962. *Phenomenology of Perception*, C. Smith (trans.). London: Routledge.

Merleau-Ponty, M. 1964. *The Primacy of Perception*. Evanston, IL: Northwestern University Press.

Montaigne, M. de [1580] 1993. "To Philosophize is to Learn How to Die". In *The Essays: A Selection*, M. A. Screech (ed.), 17–36. Harmondsworth: Penguin.

Moran, D. 2000. *Introduction to Phenomenology*. London: Routledge.

Mulhall, S. 1996. *Routledge Philosophy Guidebook to Heidegger and* Being and Time. London: Routledge.

Nabokov, V. [1935] 1959. *Invitation to a Beheading*. London: Weidenfeld & Nicolson.

Nagel, T. [1979] 1993. "Death". In *The Metaphysics of Death*, J. M. Fischer (ed.), 61–9. Stanford, CA: Stanford University Press.

Nietzsche, F. 2004. "Why I am so Wise". In *Ecce Homo*, 7–18. Oxford: Oxford University Press.

Nordenfelt, L. 1987. *On the Nature of Health*. Dordrecht: D. Reidel.

Nussbaum M. 1994. *The Therapy of Desire*. Princeton, NJ: Princeton University Press.

Plato 1942. *Phaedo* in *Plato: Five Dialogues*, H. Cary (trans.). London: Everyman.

Polt, R. 1998. *Heidegger: An Introduction*. London: UCL Press.

Rilke, R. M. 1982. *The Selected Poetry*, S. Mitchell (ed. and trans.). New York: Random House.

Sontag, S. 1978. *Illness as Metaphor; AIDS and its Metaphors*. Harmondsworth: Penguin.

Svenaeus, F. 2000. *The Hermeneutics of Medicine and the Phenomenology of Health*. Dordrecht: Kluwer.

Toombs, K. 1988. "Illness and the Paradigm of the Lived Body". *Theoretical Medicine* 9: 201–26.

Toombs, K. 1995. "The Lived Experience of Disability". *Human Studies* 18: 9–23.

Toombs, K. 1999. *The Meaning of Illness: A Phenomenological Account of the Different Perspectives of Physician and Patient*. New York: Springer.

Toombs, K. 2001. *Handbook of Phenomenology and Medicine*. New York: Springer.

Warren, J. 2004. *Facing Death: Epicurus and His Critics*. Oxford: Clarendon Press.

Williams, S. J. 2003. *Medicine and the Body*. London: Sage.

Index